NATIONS FROM IMITATIONS TO INNOVATIONS

THE HISTORY OF INNOVATION & TECHNOLOGY DEVELOPMENT IN KOREA & JAPAN

BY
PROF. MOHAMMED AHMAD S. AL-SHAMSI

I0044957

Nations from Imitations to Innovations
The history of innovations and technology development
in Korea & japan

Copyright © by Mohammed Ahmad S. Al-Shamsi

All rights reserved. No part of this book may be reproduced or used
in any manner without written permission of the copyright owner
except for the use of quotations in a book review.

Part of the content is licensed under a creative common
attribution 4.0. International (cc by 4.0) license.

1st Edition

Published Date: 2022

ISBN: 978-1-7346287-8-4

To My Father,
Oh God, have mercy on my father
and make his grave kindergarten
from the garden of paradise

ACKNOWLEDGMENT

To all those who supported me in the process of writing this book during the last two years. I am grateful to Prof. Ahmad Mohammed Al-abdulgader for his valuable and unlimited support. I am also grateful for King Abdulaziz City for Science & Technology (KACST).

TABLE OF CONTENTS

PART TWO
Imitation and Innovation in Korea

PREFACE

In our previous book, *Innovation and Imitation for Nations*, we presented the experiences of innovative nations since the dawn of history passing by medieval Europe and the United States of America before independence. We referred to the role of copycatting and imitation as a stage to which countries resorted to reaching innovation. We also presented innovations immortalized for thousands of years and how they were transferred from one country to another using simulation, copying, and imitation. We analyzed economic theories on the role of innovation in countries' economies.

This book is dedicated to Japan and Korea, which have had an impressive economic performance that gained them a well-deserved place among industrialized countries. Their advances in robotics, electronics, auto, and ship industries demonstrate industrial maturity and innovative content. However, today's view of these countries differs from our ancestors' who looked to Japan and Korea when they started manufacturing as producers of low-quality products imitated and copied from Western technologies.

To know how they reached their position today, we had to get to the depth of the process by researching the roots of these innovations over three thousand years, identifying specific innovations, and tracing their transmission or copying to these two countries.

Although the two countries share their approach to copying and imitation as a starting point, the technology transfer and development model differs between Japan and South Korea. After other countries refused to license and transfer modern technologies, Korea resorted to its national research system to add innovative content to its counterfeit products. As for Japan, it adopted several patterns over thousands of years to reach innovation.

This book summarizes the experience of the two countries, from not having the technology to innovation and economic progress, which led to the rise in per capita income, prosperity of urbanization, and the recovery of markets in both countries. These experiences may be rich and valuable, especially for today's developing countries. They are not pure theories but rather a successive series of procedures undertaken by the government of these countries. It was not a matter of chance, and their economic prosperity was not due to the sudden discovery of natural wealth.

Author
3/3/2022
Riyadh

PART ONE

IMITATION AND INNOVATION IN JAPAN

CHAPTER ONE

ANALYSIS OF JAPANESE INNOVATION HISTORY FROM THE THIRD CENTURY BC TO THE SEVENTEENTH CENTURY CE

Japan is a unique model when studying innovation and imitation in countries and nations, with its history of ups and downs, repeated introversion, and opening up to others. Unlike many other countries, we cannot set a date for its transition to imitation and its shift to innovation. Our analysis of the Japanese innovation history showed five historical periods in which the Japanese nation turned to innovation (and imitation). The first period was 2300 years ago during the Yayoi reign; the second in the seventh century CE, known as the Taika reforms; the third from the sixteenth century CE since the Nobunaga era and the extension of the Tokugawa rule; the fourth from 1853 CE to the end of World War II in 1945 CE; and the fifth period after World War II.

In each of these periods, Japan turned into an innovative country. The economy and trade flourished, accompanied by superiority in military armament, especially in the second, third, and fourth periods. Many scholars of innovation in Japan missed the first three periods. They considered the innovation in Japan only in the fourth and fifth periods, which started with the Meiji era and after World War II up to the contemporary period. This caused a lack of understanding of how the Japanese nation transformed into an innovative nation and losing sight of the historical accumulations that contributed to the nation's transformation. This view is superficial and deficient, as the one who looks at the tip of the iceberg while ignoring the unseen larger part underwater. To avoid the mistakes of other scholars, we had to reach the historical beginnings of innovation before we turn to the contemporary situation.

Historically, Japan did not invent paper, silk, or porcelain. These innovations existed thousands of years ago among the people west of Japan. Still, the Japanese successfully transferred, copied, and imitated the relevant techniques until Japan manufactured and exported these products. These techniques contributed to the economic renaissance of Japanese empires throughout the ages.

FIRST PERIOD: INNOVATION AND IMITATION DURING THE YAYOI REIGN (300 BC)

The first innovation period in the history of Japan was during the Yayoi era (300 BC-300 CE). Before that era, Japan relied on food collection by hunting and fishing, mainly using stone and clay tools. Japan had not known Agriculture, mining, and settlement. When agriculture entered Japan in the Jomon era, it was accompanied by a tremendous technological development brought from the countries west of Japan (to be more precise, it was a technology transfer process from Korea and China) (1).

Some historical sources refer to the ice movements in the fourth century BC. The resulting climate changes prompted the peoples inhabiting the southern regions of the Korean Peninsula (Paekche region) to cross the strait to Japan. The immigrants brought in the techniques found in Korea and China, as agriculture was known in China since the fourth millennium BC, more than 3000 years before it entered Japan. Iron and bronze tools used for cultivation were also part of such technology transfer, accompanied by techniques for preserving and storing rice, such as constructing underground stores to preserve carbonated rice. The rice was exposed to low heat and ventilation to acquire the carbonated husk for preservation (2, 3).

The first metal tools were brought to Japan via the Korean Peninsula. They were not made in Japan as neither the techniques of high-temperature furnaces nor of extracting metals from the earth had yet been adapted/ localized (imitated and copied) in Japan. Antiquities found in Japanese tombs, including bronze daggers, mirrors, and arrowheads, were of Korean origin, the same as those found in Korean tombs hundreds of years before. This indicates their origin was the Korean Peninsula. Metal spinning and weaving tools, such as spindle whorls and loom dating back to the Yayoi period, were imported from China and Korea. Stone ruins contained drawings of hundreds of Japanese elites dating back to the Yayoi period wearing Korean costumes and clothes. This indicates the transfer of the necessary tools for agriculture and tissues, clothing, and consumer taste in the beginning (3).

Later in the Yayoi era, the Japanese began to copy and imitate these techniques in agriculture, weaving, and weapon, followed by making modifications to the original products to suit the taste of the Japanese consumer, such as an increase in the sizes, thickness, and lengths of swords and shields, as well as in clothes and costumes.

The Japanese had shown their innovative ability by making marginal modifications and additions to the original products until they became fundamental and different. The most prominent techniques transferred from China included digging channels, the hilt ring, and silk spinning. The most prominent items from Korea included agricultural or religious metal tools, kilns, bells, and fortifications (3, 4).

This shows that the first period of innovation in Japan, which began in 300 BC, was triggered by the arrival of Korean immigrants to some Japanese shores. These crafts and tools spread to the rest of Japan, first by copycatting, imitation, and replication, and then adding local content and a distinctive consumer taste that took several generations to develop. The technological shock caused by Korean immigrants made the Japanese feel the immigrants' technological superiority. Their imitation of this superiority led to modifications, innovation, and development.

⤳

SECOND PERIOD: INNOVATION AND COPYCATTING IN THE TAIKA ERA (SEVENTH CENTURY CE)

In the seventh century CE, Japan had a technological shock after its defeat in the Battle of Pakchon, also known as the Battle of Bochuan, in Korea at the hands of Imperial China. More than ten thousand Japanese soldiers were killed, many more fell into captivity, and 400 Japanese ships were sunk. This shock was a turning point in Japan's history; the Japanese knew their capabilities were inferior to those of China, which prompted them to carry out many radical organizational and institutional changes *a la* the Chinese model. The Japanese believed China excelled because it had innovative systems, institutions, and technologies. We need to go back to Japan in the fifth and sixth centuries CE to analyze this shock. The governmental regime extended from the kofun period, where clan chieftains ruled cities and villages autonomously and owed allegiance to the Japanese emperor (5).

When the Tang Dynasty (618-906 CE) inherited the Sui Dynasty (518-618 CE) in China in 618 CE, the Tang Dynasty needed thirty years to consolidate its rule internally before kicking off its efforts to unify the Korean Peninsula. At that time, the Korean Peninsula was divided into three kingdoms: Koguryo/ Goguryeo Kingdom, north of the Korean Peninsula on the border of China, Silla/ Shilla Kingdom, west center, and the Paekche Kingdom, southeast, whose waters overlooked Japan (about 50 km by sea) (6, 7).

The campaigns of the Chinese emperor against the neigboring Koguryo Kingdom began in the forties of the seventh century CE. When the first three campaigns failed, China pursued a different policy. It allied with the Silla Kingdom, which bordered the Koguryo Kingdom to the south and the Paekche

Kingdom to the north. In this way, China get around the Kogu-ryo Kingdom from the north and south and occupied its capital Pyonyang. Then, the Silla and Tang attacked the Paekche King-dom and occupied its capital Ungjin. The Paekche Kingdom resorted to the Emperor of Japan [Emperor Tenji (661-671 CE), known for killing the leader of the Soga Clan when he was a prince in 645 CE] to request military assistance. Emperor Tenji sent troops from 661 to 663 CE to the tune of 42 thousand Japanese soldiers. China sent 7,200 Chinese soldiers to the Silla Kingdom. In 663 CE, the Battle of the Paekchon River took place. Japan was decisively defeated, and China unified the Korean Peninsula under the rule of the Silla Kingdom, which was subordinate and loyal to the Tang Dynasty in China (6, 7).

After this incident, the Japanese recognized China's techno-logical superiority and imitated it. It began with the techniques of forts and castles on the shores of Japan and then with a series of systematic and institutional reforms and the abolition of all privileges of the old tribal groups as part of the so-called Taika Reforms (8, 9).

Some studies (10) have proven that the fundamental Taika Reforms did not start with the rule of Emperor Kōtoku in 645 CE but rather after the shocking catastrophic defeat during Tenji's rule rendering the historians' claims otherwise invalid. We support this rationale in our book, as significant changes require an extreme shock. Otherwise, societies would not accept change, and leaders would not spur it. Here, we are not interested in reforms of ranks and titles, the distribution of land, and the establishment of institutions as much as the shift to technological change, which is to imitate and emulate the superiors after the technological shock, as was the situation in Japan after the defeat of 663 CE.

A notable change in terms of organization and administration included the re-division of Japan into 66 provinces and the appointment of governors, dividing these provinces into 592 regions and the regions into smaller units per the size of the villages in each. In a later book, we will present the innovations in laws and regulations. As for this book, its primary focus is technological innovations. Unfortunately, most historians who mention the Taika era focus on administrative reforms while overlooking the technological aspect.

〜〜〜

CHAPTER TWO

CHAPTER TWO

INNOVATION OF MECHANICAL WATCHES AND FIREARMS IN THE NOBUNAGA ERA IN THE SIXTEENTH CENTURY

THIRD PERIOD: INNOVATION IN THE NOBUNAGA ERA (SIXTEENTH CENTURY)

In the sixteenth century CE, Japan began copycatting and imitating a set of technologies that reached its shores and then moved to developing and innovating these imported technologies during the following two hundred years. Our focus here is the techniques of mechanical watches, mechanical puppets, and rifles during Japan's third period of copycatting and innovation.

We will start with a historical overview of the watches developed by humans up to the mechanical watches, which later became an industry and an innovative product in Japan. So far, Japan feeds and develops this industry worldwide; the number of watches manufactured in Japan in 2019 was 66 million pieces, with sales exceeding US$ 2.6 billion (11).

Back to the history of the watch innovation, let's see how it arrived in Japan first as a commodity and then as an industry. In 1500 BC, the ancient Egyptians invented sundials, which operated only during the day to use the sun's movement to tell the time. Then, in 1400 BC, the Egyptians invented the clepsydra, a water clock, a timepiece by which time was measured by the regulated flow of liquid into or out from a vessel to overcome the shortcomings of shadow and cloudy hours. It used the division of the day into 24 hours (12-14).

The Greeks invented the sand clock in 500 BC, adopting the concept of dividing the clock into 60 parts (minutes) from the Babylonians who had borrowed it from the Sumerians who had developed this concept around 2000 BC. Then, the Chinese invented the candle clock, a thin candle with consistently spaced marking that, when burned, indicates the passage of periods in 520 CE (15).

Then, humans reached the mechanical escapement clocks. The first mechanical clocks installed on a public tower in Italy in 1315 CE were mentioned in a novel published in 1321 CE.

They were also found in 1365 CE in other Italian cities made by Giovanni Dondi dell'Orologio. Then, this industry moved from Italy to England to in 1386 CE to know the time for religious purposes and prayers. The oldest surviving piece of a mechanical clock is in Salisbury Cathedral. These mechanical clocks were installed on buildings and not individual watches. This technique spread for nearly 200 years in Europe until 1505 CE, when the German Peter Henlein invented the first watch in Nuremberg placed in a pocket or attached to a necklace (it was not a wristwatch). It was a closed ball connected to a chain. Wristwatches appeared in 1810 CE when Abraham-Louis Breguet, founder of the Breguet brand, made a wristwatch for the Queen of Nablus. Patek Philippe made another wristwatch for the Hungarian Countess Koscowicz in Switzerland in 1868 CE. The price of this masterpiece exceeds US$30 million today (16, 17).

How did the watch industry move to Japan? How did Japan become an innovative country in this field? The focus here is not on non-mechanical watches, as Emperor Tenchi made a water clock in 660 CE (or 671 CE, according to other sources) as mentioned in *The Chronicles of Japan* (*Niho*n Shoki), the second-oldest book of classical Japanese history.

In 1551 CE, after the Spanish and Portuguese discovered the sea route to Japan, a Spanish missionary, Francis Xavier, gifted a mechanical watch to a Japanese feudal lord named Ōuchi Yoshitaka, the daimyo i.e., ruler, of Suō Province. Yoshitaka liked it and kept it to show his visitors. There were frequent cases of missionaries giving gifts to Japanese feudal lords to entice them into Christianity, or at least allow them to proselytize in their areas of influence actively; they partially succeeded. Clocks were also presented to Oda Nobunaga in 1569 CE and Toyotomi Hideyoshi in 1591 CE. The Pope in Rome sent a gift to Tokugawa Ieyasu in 1606 CE when ruling Japan as shōgun, (Commander-in-Chief of the Expeditionary Force Against the Barbarians), the title of the military dictators of Japan (18).

After the high demand for watches by the aristocracy in Japan, merchants began to import them to sell them to the elite. After 47 years from their entry into Japan, Tsuda Suke Zamon, a Japanese repairer of European watches, made the first mechanical watch in 1598 CE, resembling those imported from the Spaniards and the Portuguese.

The watch industry spread throughout Japan. However, the difference between the Japanese and the European calendar hindered that spread. The Japanese had to invent the so-called pillow clocks, or wadokei, a uniquely Japanese style that spread and flourished locally to overcome this obstacle. In 1796 CE, Hosokawa Hawzo wrote a guideline book explaining how to make mechanical watches, which immensely helped the emergence of Japanese artisans and watchmakers. The spread of this watch model continued until 1872 CE, when Emperor Meiji issued a decree to change the Japanese calendar to a solar calendar like the European calendar (18, 19).

The Japanese watchmakers developed watchmaking and gear technology until they introduced the gears used in mechanical watch movements to transform fixed dolls into mechanically moving dolls. In the seventeenth century, mechanical puppets, called *karakuri* ("to pull, stretch, and move a thread"), appeared. They consisted of a doll's head in a robe and two hands, and inside it, a group of interconnected gears to move the doll. These dolls were popular with the aristocracy first because of their high cost. Tanaka Hisashige, called Thomas Edison of Japan, was a pioneer of the mechanical doll industry in Japan. He also later contributed to the introduction of many Japanese industries, which we will mention in another section when we analyze the technological transformation during the reign of Emperor Meiji (18, 20, 21).

Back to the analysis of this innovative technology, the Japanese could not attain that until sometime after the mechanical watch industry entered Japan. Accordingly, importation, then imitation, copycatting, and simulation led the Japanese to innovate in the watch industry first and produce other industries later that had the exact working mechanism after a short period (less than two centuries).

—❦—

INNOVATION AND IMITATION OF FIREARM TECHNIQUES IN THE NOBUNAGA ERA IN THE SIXTEENTH CENTURY

This part tracks and analyzes the techniques of weapons transferred to Japan and copied and then developed over time in the sixteenth century, especially the rifle and artillery.

In the first half of the sixteenth century, the common weapon in Japan was the sword and spear. Individual firearms, such as rifles and pistols, were unknown in Japan. On August 25, 1543, a Chinese ship landed on the Tanegashima Island in southern Japan, carrying 100-s crew of Chinese sailors and three Portuguese sailors whose ship was wrecked only to be rescued by that Chinese ship. They had two Ottoman arquebus/musket rifles. When a Japanese saw a Portuguese sailor hunting ducks with a rifle, he informed the governor of the island, who in turn offered 1,000 gold coins for each rifle. The governor then handed one of the two rifles to his blacksmith and swordsmaker and asked him to replicate the weapon. The blacksmith made ten guns for the governor within a year. Then, this industry spread in Japan. In 1556 CE, (that is, nearly a decade later) the number of rifles in Japan exceeded 300,000. During this decade, tribal leaders and regional princes raced to attract skilled blacksmiths and poured money on them to make rifles for their forces. Initially, the Kyushu Island was the center of the firearms industry before its spread throughout Japan. The cities of Sakai, Yokkaichi, and Kunitomo became famous hubs for the manufacture of firearms in Japan (22-25).

It is worth mentioning that the awareness of the tribal leaders of the benefits of this type of weapon contributed to such a dissemination of the rifle. In 1549, Oda Nobunaga, the leader of the Oda clan (a young man in his twenties at the time), ordered blacksmiths in the Kunitomo region to manufacture 500 rifles. In 1560 CE, he introduced the rifles to his forces, which used three thousand rifles to defeat the Takedos forces at the famous Nagashino Battle. Nobunaga became the most potent tribal chief in Japan and the first of the Three Unifiers of Japan. His successors, Toyotomi Hideyoshi and Tokugawa Ieyasu followed his path and strategic methodology in supporting the rifle until they united all Japan after the decisive Battle of Sekigahara in 1600 by Tokugawa. In 1592, Hideyoshi invaded and occupied the Korean Peninsula in 11 days (and in another version 21 days) thanks to the arming of the entire Japanese army with rifles. They also defeated the Chinese forces that came to the aid of the Koreans. The rifle was not present in Korea at the time. The Koreans and Chinese could not liberate Korea until after the rifle-making technology was adapted on the Korean Peninsula. Once the local industry of the weapon spread, they expelled the Japanese invaders (or conquerors) in 1598. In other words, the copycatting and simulation of the Japanese secured the liberation of the Korean Peninsula (22-26).

During the war of unification of Japan, the challenges Nobunaga faced were only with those who adopted the armament and manufacture of rifles extensively. For example, Hōjō Ujinao was the last tribal chief who resisted to the end thanks to the number of weapons and ammunition he made. It was said that each door of his Odawara Castle had a cannon and three rifles, an indication of the prevalence of weapons and the adaptation of this technology in the lands under his influence (22).

We cannot fail to mention here that the rifle in Japan in 1600 was not the same as the one that entered the island in 1543. The Japanese developed this weapon in several aspects, including size, weight, decoration, nozzle size, and ammunition weight. In other words, the Japanese innovated, thus rendering Japan an advanced country in this industry at the beginning of the seventeenth century.

As for artillery, although some sources refer to its entry into Japan since the thirteenth century CE during the Mongol invasion, it did not move as an industry until the sixteenth century. In 1551, Otomo Yoshizumi, a Japanese daimyō of the Sengoku period, was briefed by two spies on the Portuguese cannon, a gift the King of Portugal gave to the tribal ruler. Attempts were made to replicate and manufacture the cannon in the provinces under his control (26, 27).

In 1558, an attack on the province was repelled by cannon shots, thus indicating the transfer and installation of technology for defense in less than seven years from the arrival of the first modern cannon to Japan. In 1571, Oda Nobunaga adopted artillery and assigned blacksmiths to manufacture a cannon carrying a shell weighing 750 grams. Then he used a group of cannons in 1578 in the Battle of Noguchi to break down forts and castles. The success of these improvements led to the spread of cannon-making technology in Japan (22, 27, 28).

In 1582, Nobunaga's successor, Toyotomi Hideyoshi, captured the fortified Kanki Castle with the help of cannons. In 1615, Tokugawa Ieyasu, using 300 cannons, demolished Osaka Castle. However, the cannons manufactured in Japan did not outperform the European ones for several reasons, foremost of which was the nature of the Japanese land, which had few castles and fortresses compared to Europe. Therefore, Japan kept importing and manufacturing cannons for 60-70 years since 1551. The Dutch were the leading suppliers of cannons to Japan. The cannons were installed on ships and used in several battles, including the Battle of Nagashima and Kizugawaguchi in the 1670s (27, 28).

<hr />

LAWS AND REGULATIONS FOR ABANDONMENT OF FIREARM TECHNOLOGIES

In his book *Giving Up the Gun: Japan's Reversion to the Sword*, in 1979, Noel Perrin explains how the industrialization and technological development of weapons, such as guns, cannons, and castles, stopped from the beginning of the seventeenth century to the middle of the nineteenth century, that is, for more than 200 years, a reversion that generated the tremendous technological shock upon the Arrival of the American Black Ships, as called by the Japanese, in 1853 in The Perry Expedition (23).

Following the victory in the 1600 battle, led by Tokugawa, he unified Japan. He tried to prevent weapons manufacture inside Japan or importing them and issued laws to make only one city manufacture weapons under his direct supervision. To obtain a weapon, the approval must be stamped by him. Within 30 years, Japan was engulfed in complete isolation. Tokugawa even passed a law to execute any Japanese who travel abroad or return from abroad. In 1639, after the Christian missionaries' revolution, foreign relations were restricted, and the English, Portuguese, and Spaniards were expelled from Japan. Only the Dutch remained on a small island where they had an embassy while not permitted to enter the other islands of Japan. This situation remained until the Americans arrived in 1853 CE (28).

To shed light on the transfer of rifle and artillery technology to Japan and then to abandon the development in this field through the study of laws and regulations, it is noticeable that regulations and laws did not play any role in technology transfer, copycatting, imitation, and innovation in the armament that followed. The tribal leaders' need for military superiority generated intense competition for the manufacture of rifles and cannons and the development of these weapons. The competition between local leaders has led to the injection of money into this sector, which has generated an arms market and the establishment of a number of cities as hubs for the manufacture of weapons. In other words, the lack of legislation was a catalyst for this industry.

As for abandoning development and stopping the manufacture of weapons, it was not the desire of the local leaders (or the lack of demand for them in the markets), but instead enacting a series of regulations and laws to ensure the non-proliferation of weapons in a bid to control the country after unity. These systems included the decision issued in 1607 to limit firearms manufacturing only in one city (Nagahama) and the decision to prevent the construction of ships capable of crossing the ocean in 1650 and only manufacturing fishing vessels. Cannons were made to be installed on ocean-going ships. When the manufacture of these ships was discontinued, the demand for cannons fell off (23).

The decision of 1607 restricting the weapons manufacture to one city under the supervision of the central administration and the decision that that stipulated prior approval of the central ruler complicated the process to obstruct and govern the issuance of regular licenses for arms sale and manufacture. A decision was also issued preventing arms makers (blacksmiths) from moving and restricting their stay only in the city designated for making weapons; otherwise, they should abandon the profession. In 1632, a commission was established to control the spread of arms and ammunition, entrusted with issuing a license for arms and ammunition and regulating their movement within the country. In 1635, Sakoku issued a decision of almost complete isolation, banning travel and issuing the death penalty to any Japanese traveling or returning from abroad. He also issued a subsequent decision banning import from foreigners, restricting that, and limiting it to only China and the Netherlands, and to one island where they were subject to permanent control from the central government (23).

Thus., transferring the military industry to Japan was not based on laws but rather a spontaneous desire to obtain a commodity to prevail in battles. As for stopping development and innovation in this field, it did not come naturally, but rather through a series of laws that took nearly 40 years to form until the weapon was entirely under control. The fighters returned to using swords and spears, development and innovation stalled. It is worth noting that the amount of development in Japan in the arms industry was significant and highly competitive with that in Europe at that time. A small fleet carrying modern cannons was manufactured. Although these prohibitive measures ensured stability and internal security in Japan for nearly 250 years until 1853, they caused a severe shock to Japan after its contact with the world after a prolonged period of isolation. The next chapter explains how such a shock drove Japanese people to copycatting and innovation again.

CHAPTER THREE

CHAPTER THREE

INNOVATION AND IMITATION OF STEAMSHIPS IN JAPAN SINCE THE NINETEENTH CENTURY

FOURTH PERIOD: INNOVATION AND COPYCATTING OF STEAMSHIPS IN JAPAN (NINETEENTH CENTURY)

Japan's semi-isolation from the outside world made its military technologies outdated in the nineteenth century, as Japan was still reliant on swords and spears in combat. Meanwhile, the USA began to rely on steamships that used coal as fuel. When voyages flourished from the USA to China and the east across the Pacific Ocean, there were no coaling stations during the long voyages, which hindered the movement of smaller steamships that could not load large quantities of coal. Moreover, ships carried less cargo because they were harnessed to load coal fuel (29-31).

In addition, the coaling stations required maintenance, advanced mechanical equipment, and intensive labor to recharge steamships. The US president wanted to open a coaling station in Japan, an island in the middle of the Pacific Ocean between the USA and China.

The whaling industry also played a significant role in raising interest in Japan. The Pacific Ocean whale trade was booming, reaching its peak in 1846. The number of coal-fueled steamships was 735 crewed by about 70 thousand. Accordingly, coaling stations in Japan and the Philippines were necessary to serve this industry. An official mission was sent in 1846 with two ships led by James Biddle. One of the ships contained 72 cannons, but the request to establish a coaling station in Japan was refused (32).

The whaling sector was important as it was one of the five largest sectors of the US economy and generated more than US$ 10 million (in 1880) of USA gross domestic production (GDP). The number of whaling steamships in the USA exceeded all the whaling ships around the world combined at that time. Based on these reasons, in 1853, the thirteenth US President Millard Fillmore (1850-1853) sent Commodore Mathew C. Perry with four small steamships to Tokyo Bay to force the Japanese to sign a treaty with the Americans for the reasons mentioned above. Trade with Japan was not a goal of the original agreement. Perry directed the steamships' artillery toward the city and fired several shots in the air to make sure none would get in the way to the port (he later said this was to celebrate the US Independence Day). The ships arrived on July 8, 1853 without resistance amid the amazement and astonishment of the Japanese. He landed on July 14, handed the message to the Japanese emperor, and promised to return next year to receive the reply. This forced the Japanese to sign the Kanagawa Treaty on March 31, 1854. Again, trade exchange was not important to the US side in this treaty, and a trade agreement was not drawn up for it until 1858 (29-31, 33, 34).

What made the Japanese emperor (or the Tokugawa government in control then) accept the 1854 treaty and not resist the steamships in 1853 even though the Dutch informed him that the US ships were on the way.

The tribal leaders convinced the emperor that Japan did not have land or sea weapons equivalent to Americans'. Everyone in Japan felt the sizeable technological gap between the US and Japanese naval forces, which prompted them to accept the 1854 treaty on unfair and unequal terms.

This technological difference shock prompted Japan to enter into a full-fledged era of copycatting, imitation, and simulation in various fields for more than half a century before embarking on local innovation for these imported technologies. The next part highlights a group of technologies and industries that Japan began to copy up to the innovation stage, most notably steamships, cotton, and motorcycles.

~

JAPAN'S TECHNOLOGICAL SHOCK IN 1953

The technological shock that Japan received was due to the four US ships, three of which were steam- and coal-fueled and equipped with advanced cannons as per the specifications in the table below. These ships made Japan acknowledge the wide technological gap between it and the owners of these ships and decide to break the barrier of isolation that lasted nearly two hundred and fifty years and allow these ships to enter the port unresisted. Japan's political leaders admitted the futility of engaging in a battle with those who possess advanced technologies. Steamships fascinated Japan and made it acknowledge the superiority of its opponents.

Table: Specifications of the four ships that entered Tokyo Bay in 1853, commanded by Commodore Mathew C. Perry.

Ship Name	Engine Type	Speed (Knock)	Weight (Ton)	Dimen- sions (meter)	Year of Manu- facture
USS Mississippi	Steam	8	3272	70 x 12	1841m
USS Susquehanna	Steam	10	2489	78 x 14	1850m
USS Powhatan	Steam	11	3825	77 x 14	1850m
USS Lexington	Paddle	-	-	39 x 10	1825m

In that period, the manufacture of ocean-going ships was banned. Only wooden fishing ships without combat cannons were allowed. How could Japan become the second-largest global ship manufacturer in the 1930s and acquire the largest market share for selling ships for three consecutive decades from 1969 to 1999? All this started in 1853 when Japan did not have a single steamship, a single iron ship, or a single ship capable of crossing the ocean. This part of the chapter explains the stages and steps that Japan adopted over a hundred years to become the first shipbuilding nation worldwide.

⟿

INNOVATION AND IMITATION IN SHIPBUILDING DURING THE TOKUGAWA ERA

To dissipate some historical confusion in technological transformation and innovation in Japan, we have to point out that the technological transformation did not start during the reign of Emperor Meiji (1867-1869), as many historians of Japan's history wrongly indicate. Instead, it started with the arrival of the four US ships in 1853. At that time, the system of government in Japan stipulated that the emperor was just a symbol with no authority, an army, or resources. He could not make public affairs decisions. Instead, all power was in the hands of tribal leaders from the era of Tokugawa Ieyasu until 1869 (35). In 1869, the emperor won the Boshin War and then took power autocratically in cooperation with the occupying foreign forces. Accordingly, from 1853 to 1869 (about 16 years after the arrival of the US ships in Tokyo Bay), all the decisions to transfer modern technology from the West to Japan were by the Shōgun (the de facto ruler of Japan). This was the first stage for technology transfer and adaptation in Japan in the modern era, followed by the stage of Emperor Meiji, starting in 1869, when he spread his hegemony over Japan (36, 37).

The Shōgun quickly responded to bridge the technological gap between Japan and the West after receiving the technological shock. A decision was issued to cancel the ban on the manufacture of the ocean-going ships. All tribal chiefs and provincial governors were instructed to start making steamships like those owned by the Americans and Europeans seen docking in Japan's ports. Fourteen local governors responded immediately in the same year (1853) and before even signing the treaty with the US and began copycatting the US steamships. We may summarize five strategic directions in the first stage of technology transfer (1853-1869) taken by Japan to bridge the technological gap in the shipbuilding industry. The first was the import of steamships from more than one western country. The second was the construction of local shipyards in different provinces. The third was establishing maritime training centers in 1855, such as the Nagasaki Naval Training Center and the Torpedo Training Center in Yokosuka in cooperation with the French (later). The fourth was sending Japanese trainees to study in Western countries. In 1862, fifteen Japanese trainees were sent for three years to Rotterdam in the Netherlands. The fifth direction was the establishment of the government-subsidized companies specialized in heavy industries, such as the IHI Corporation in 1853 in Ishikawajima (38-41).

In terms of import, the import of steamships began immediately after the 1853 incident. Building on the long relations between the Dutch and the Japanese, dating back to 1641, after the expulsion of all foreigners from Japan, Japan imported the

Kanko Maru steamship from the Netherlands in 1855. It was a gift from King William III, with a maximum tonnage of 781 tons and dimensions of 66 meters in length and 9.1 meters in width, and had six cannons. William III sent 22 Dutch sailors to operate it.

Japan also imported Kanrin Maru in 1857, which had a tonnage of 300 tons, dimensions of 50 meters in length and 7.3 meters in width, and contained 12 cannons at a speed of 6 knots. These two Dutch ships were wooden and steam-powered by providing them with coal.

An iron steamship called the Kotetsu (literally "Ironclad," later renamed Azuma, "East") was imported from France in 1867. The French were the first to manufacture iron-armored ships in 1859 when they launched the Gloria with a tonnage of 5,630 tons and a speed of 13 knots. The ship imported to Japan was a simulation of this ship.

Japan also imported a series of English steamships manufactured in Aberdeen in the 1860s, including the steamships Josho Maru, Hosho Maru, Kagoshima, and Banryu, manufactured in England in 1858, and Chogei, manufactured in Glasgow in 1866. Also, a gift was received from the King of Prussia in 1864, a steamship called Kaiten, and another, called Shinsoka, from the US.

As for shipyards, the Shogunate contracted with the French to build large Western-style docks in Yokosuka and Nagasaki. The first phase before the rule of Emperor Meiji in 1869 had five shipyards. The shipyard in Nagasaki, which began operating in 1857, was the first modern shipyard in Japan. It was dedicated to repairing minor breakdowns and manufacturing ship engines. As for the Yokosuka shipyard, it was the largest project at that time, as it took seven years to build (1864-1871) at the cost of US$2.4 million. It was staffed by forty French engineers and technicians and 2,000 Japanese employees.

In terms of the local industry of modern steamships, the ruler and leader of the Mito clan manufactured the Asahi Maru ship in 1856, which took two years to build, with a tonnage of 750 tons, dimensions of 426 meters in length and 9.1 meters in width, containing 20 cannons (ten on each side).

The staff of the port administration in Uraga (the only port dedicated to examining foreign ships coming from the ocean) made the Hoo Maru with a tonnage of 600 tons and dimensions of 36 meters long and 9 meters wide and ten cannons (five on each side).

The governor of Satsuma Province built the steamship Shohei Maru in 1854, with dimensions of 31 meters long and 7.3 meters wide and ten cannons (five on each side).

The governor of Chosho Province built a steamship in 1860. It is no secret that the locally manufactured steamships between 1853 and 1869 were of poor quality compared to the imported ones. When the Meiji rule began, it focused more on importing, with a decline in the local industry, due to the difference in quality.

On the corporate front, IHI Corporation manufactured the Asahi Maru mentioned above to the leader of the Mito clan in 1856, three years after its inception. In 1863, the company manufactured the Chiyodagata gunboat, which took 26 months to build, with a tonnage of 140 tons, a speed of 5 knots, and dimensions of 31 meters in length and 4.8 in width.

During the Boshin War (1868-1869), the famous naval Battle of Hakodate took place between the forces loyal to the emperor and the tribal government. In this battle, 13 Japanese steamships, eight ships loyal to the emperor and five to the Shogun, engaged. The emperor's forces prevailed after sinking two ships and seizing the other three thanks to the emperor's foreign support as the legitimate ruler of Japan and a desire to continue trading contracts with him after victory (37, 42).

Japan entered two foreign wars, one with the Chinese and the other with the Russians, and steamships played a role in Japan's progress in battles and winning both wars. We will provide more detail on the Japanese Navy in the two wars later in the book.

IMITATION AND INNOVATION IN SHIPBUILDING DURING EMPEROR MEIJI'S REIGN

At the outset of his era, Emperor Meiji launched an ambitious plan in 1870 to build 200 steamships in Japan, but the resources needed were not enough, hence the quick abortion of the program. Emperor Meiji decided to make the British Navy the appropriate model for adoption in Japan in 1870. In that year (1870), Japan manufactured two local steamships with a total tonnage of 57 tons, compared to 382 steamships made in Britain in the same year with more than 220 thousand tons. Britain was the largest manufacturer of steamships globally. Japan used Britain as a model for imitation and copycatting. Accordingly, a British mission (Douglas Mission) was sent to lay the foundations for the fledgling Imperial Japanese Navy (1873-1879) (43, 44).

However, imports constituted more than half of the steamships entering service during the 18 years between 1879-1896, with a total tonnage of 193 thousand tons out of 337 thousand tons (for all imported and locally manufactured steamships during this period). When the small steamships manufactured in France proved their worth in the French-Chinese conflict, Japan requested the import of 48 small French Cruise or Torpedo steamships in 1882 (45).

In 1896, the Naval Government Support system was issued and continued in place for 24 years to support the local construction of steamships and reduce imports by providing subsidies on each ship built based on its tonnage. It offered 12 yen per ton for steamships of 700 to 1,000 tons, and 20 yen per ton for steamships over 1,000 tons. This subsidy compensated the local manufacturers for the price difference in obtaining steel compared to the British shipbuilders. The raw materials that went into the manufacture of ships, the most important of which was iron, could be obtained at lower costs in Britain, which granted the British a comparative advantage over the Japanese manufacturers. The effectiveness and impact of the new subsidy decision was reflected in the intensity of production. For example, all that Japan produced during the eighteen years from 1879 to 1896 amounted to a total tonnage of 4,000 tons. After the issuance of the decision, the total tonnage of ships manufactured in Japan (1897-1902) exceeded 20 thousand tons, nearly five times the production in a shorter period. In 1898, Japan manufactured the Hitachi Maru ship with a capacity of 2000 tons. In 1907,

Mitsubishi built the Tenyo Maru ship with a capacity of 13,444 tons and a speed of 20 knots. In 1913 (before World War I), Japan had five shipyards capable of producing steamships with a carrying capacity of more than 1,000 tons: Uragu, Ishikawajima, Osaka, Kawasaki, and Nagasaki (43, 44, 46-49).

Japan reached this somewhat advanced stage despite the scarcity of raw material (iron) and its dependence on imports, as it imported more than three-quarters, 34,000 tons out of 44,000 tons, of its iron needs.

⋙

THE IMPACT OF COPYCATTING AND INNOVATION IN SHIPBUILDING ON THE SINO-JAPANESE WAR

When the Sino-Japanese War took place (July 1894-April 1895) to control Korea, the Chinese naval forces had the largest naval fleet in East Asia. The defeat of China in the war with France (1884-1885) led it to focus on the Navy and the purchase of warships from Britain and Germany. In 1888, China's naval fleet included 78 ships. This fleet included the two huge German ships Dingyuan and Zhenyuan, delivered to China at the end of 1885, equipped with an arsenal of cannons. However, the Japanese naval forces defeated the Chinese fleet in the Battle of the Yalu River, where they destroyed and captured eight of the 12 ships that participated in the battle (50, 51).

Japan had the upper hand due to easily maneuverable small torpedo ships. It had 26 ships of this type, let alone the training and capabilities of Japanese sailors and navigators. Japan could manufacture part of the fleet involved and could better address breakdowns. Japan entered this field about 15 years before China, as China did not produce its first steamship until 1869, though the first steamship reached the shores of China in 1828. Interestingly, a report was submitted to one of the observers in the Chinese ports that the British steamship at the port moved from a copper box fed with coal and released steam with no sailors pushing the oars. Another report was submitted at the end of the Opium War in 1842 to the Chinese Emperor that steamships operated with fire-fueled oars! The bottom line was that it took China more than 40 years to build a steamship, while Japan took only two years (52, 53).

Japan also contributed with its advanced naval fleet with the foreign countries that supported China in the Boxer Rebellion War (1899-1900). The Chinese revolted against the foreigners and their supporters for draining the country's resources. A local fighting group ranging from 50,000 to 100,000 gathered around the Qing Dynasty and killed more than 200 members of the foreign missions and thousands of Christian Chinese accused of being loyalists to the West (54, 55).

The Japanese Navy participated with a third of the war fleet, with 18 warships out of 50 participating ships from 12 countries. This shows the fruit of 45 years of work in the shipbuilding industry that led to the emergence of Japan as a regional power in East Asia at the time.

⌒≈⌒

THE IMPACT OF IMITATION AND INNOVATION IN SHIPBUILDING ON THE RUSSO-JAPANESE WAR

When the Russo-Japanese War took place (February 1904-September 1905), Japan had taken steps to modernize the Navy due to its importance in regional sovereignty. It increased more than 100 steamships with more than 200 thousand tons and raised the naval workforce from 15 to 40 thousand (56-58).

Japan had also begun manufacturing advanced warships (six), including the Mikasa, which took 21 months to build and entered service in 1902 with a tonnage of 15,000 tons and dimensions of 131 meters in length, 23 meters in width, and a speed of 18 knots. It was equipped with an integrated arsenal of war. Japan had also introduced six armored cruisers into service. However, Japan's domestic manufacturing capabilities were not ready to manufacture armored battleships and cruisers; most were imported from Britain (59) (56, 57).

In the Battle of Tsushima on May 27, 1905, the Mikasa battleship was at the forefront of the Japanese fleet against the Russian fleet in the Baltic. The Japanese Navy destroyed 21 of the 38 Russian ships; 4,500 Russian soldiers died, and 6,000 were captured. The losses did not exceed 120 soldiers on the Japanese side, and only three small torpedo ships were destroyed (59) (57, 58).

After Japan's victory in this naval battle, the war ended in favor of Japan thanks to the superiority of its naval fleet. Accordingly, it is clear that after Japan's introversion for 250 years and remaining in semi-isolation with the outside world, it could win three battles outside its borders after nearly 50 years of opening up to the world, thanks to the transfer and copycatting of marine technologies.

SHIPBUILDING IN JAPAN DURING AND AFTER WORLD WAR I

When World War I (1914-1918) broke out, Japan tapped its neutrality. The manufacturing capabilities of ships rose dramatically from five entities capable of manufacturing ships with a tonnage higher than 1,000 tons in 1913 to 52 in 1918. The number of shipyards increased from six in 1913 to 57 by the end of World War I, and four-fifths of these yards were for iron steamships (49).

The annual capacity to produce steamships increased from less than 60,000 tons in 1913 to more than 600,000 tons in 1919. In other words, the annual production capacity increased more than ten times. During this period, the import of steamships, which did not exceed 4 thousand tons in 1918, almost disappeared. During the war, Japan contributed to covering the international need for steamships of all kinds, and its exports of iron steamships exceeded 50%, higher than the US exports (49).

In the 1930s, with the advent of diesel engines for ships and replacing coal with diesel in Switzerland and Germany in 1924, Mitsubishi Corporation purchased the patent rights to manufacture this type of ship. In 1924, Mitsubishi Corporation and Kawasaki Corporation manufactured the first Japanese diesel-powered ships with a tonnage of 4,600 tons. In 1930, a law banning the import of ships to Japan was issued, allowing only local production, thus contributing to making Japan the second-largest ship manufacturer globally after Britain in 1937 (49).

In 1944, during World War II, the total production capacity of ships reached nearly two million tons, three-quarters of which for the naval war fleet. However, the defeat in World War II was catastrophic for the shipbuilding industry. Japan was forced to limit the maximum capacity of ships to 150 thousand tons - with many other conditions and restrictions, including limits on speed, volume, and tonnage – which was only less than 8% of the total production capacity during wartime (49).

Japan passed this stage, which lasted about 15 years until the 1960s. According to the total tonnage, Japan's production of ships became 9.5 million tons in 1962, making it the largest producer and exporter of ships in the world. Its share now exceeds half of the world's ship market. It maintained its status as the largest producer of ships globally until 1999, i.e., for nearly three and a half decades (60, 61).

Japan's experience in transferring the technology of modern shipbuilding was triggered by a technological shock and a catastrophe, followed by accepting the superiority of the other, then copycatting, imitation, simulation, and finally innovation.

This process eventually led to Japan's supremacy. Within nearly a hundred years of not owning any steamship in 1850, Japan became the most prominent global ship producer in 1962 (49, 60).

Back to the rationale of this book, Japan did not invent steamships, diesel ships, ship-mounted artillery, or iron ships. Instead, it imitated and copied others until it perfected such imitation and then worked on developing, improving, and pruning it up to the peak of innovation in this field as measured in terms of the global market share of the shipbuilding industry, the number of production units, the total annual production capacity, or even the manufacturing quality.

CHAPTER FOUR

CHAPTER FOUR

INNOVATION AND IMITATION IN THE COTTON INDUSTRY IN JAPAN SINCE THE NINETEENTH CENTURY

INNOVATION AND COPYCATTING IN JAPAN'S COTTON TEXTILE INDUSTRY

In the Tokugawa era (1603-1869), Japan was self-sufficient in the cotton industry using traditional methods and mechanical machines that did not rely on steam and coal. Japan produced cotton from its farms and manufactured it through two main production mechanisms: first, at homes, where homemakers spun cotton and sewed clothes, and second, cotton spinning and weaving centers, known as *sanchi*, which depended mainly on mechanical machines operated by several workers. Japan continued in this way for hundreds of years until it opened up and knew modern cotton products and machines through trade foreign (62).

Before the Tokugawa era, cotton was extrinsic to Japan as an industry and garment. Clothes were mostly made of local linen for ordinary people, who were forbidden to wear silk, exclusive for nobles. However, in the fifteenth century, imports of cotton products began from China, and a sizeable local market was created. The Japanese farmers began to transfer the techniques of cotton cultivation and weaving from China. From importing raw cotton and ready-made textiles from China to local agriculture and weaving of cotton textiles, this stage lasted for nearly 200 years. During the Tokugawa era, local cotton agricultural production and textile industry (during the self-inflicted isolation period) prevailed (63).

⇜

Although the raw material for the cotton fabric industry was scarce in Japan and its quality was unsuitable for advanced machines in the nineteenth century due to its high moisture content, Japan began to mechanize the cotton textile industry in 1866. It manufactured cotton weaving and spinning machines and became the largest exporter of cotton textiles globally in 1929; some sources indicated Japan exported 69% of the world's textiles then. How could Japan achieve this technological progress and economic growth in this field in less than 70 years? This chapter analyzes the historical incidents of technology transfer, tracks import and introduction of machinery into factories, how they grew, and the government's role in this (62, 64).

❧

TRACING THE ROOTS OF COPYCATTING AND IMITATION IN THE COTTON TEXTILE INDUSTRY IN JAPAN

The first modern machines for the cotton weaving industry were imported in 1866 from Britain to the Kagoshima Spinning Mill. The British Henry Alline made plans for the first cotton-weaving factory with modern steam machines. Then followed a group of mills that adopted the British production style and imported machinery, equipment, engineers, and consultants from Britain. The first successful project to manufacture cotton fabric in Japan, Osaka Spinning Company, even brought its bricks from Britain (62, 65, 66).

Osaka Spinning Company imitated the construction of the Broadfield and Palm Mills. From 1869 to 1880, the first models were a simulation of the mills and factories in the British city of Lancashire. The Japanese Takeo Yamabe, during the era of openness, traveled to Britain to learn about the textile industry.

He studied for a year, then worked in a mill for six months, and was impressed by the performance and steam engines. On his return after passing the guild master craftsman's exam in Britain, he decided to establish a mill in Osaka, Japan. He was appointed Executive Director of the Osaka Mill owned by Eiichi Shibusawa. After the success of this mill cotton weaving and spinning mills spread rapidly (66-68).

We must not lose sight of the efforts of the clan chiefs and independent provincial governors of the Tokugawa era before the advent of Meiji. Five local chiefs and provincial governors built modern-style mills and purchased their equipment from Britain via English merchants in Nagasaki. Satsuma Clan was among the first clans to adopt the modern cotton industry. One of the first tribal mills was founded by Godai Tomoatsu, Shinno H., and Ryoichiro Okada who founded the Futamata Textile Co. (68).

Later, the Meiji government introduced three main mechanisms to support the cotton industry in Japan (68):

1. The government provided financing to the government mills to import modern cotton spinning and weaving equipment, as in Kuwahara.

2. The government purchased and imported machinery and equipment and then sold them to the owners of

Japanese cotton mills interest-free. In the Toyoi area, the government sold equipment and machinery directly to the factories.

3. The government established and equipped mills and then leased or sold them to merchants, as in Hiroshima.

4. The government established mills to be display models to encourage small and emerging companies to emulate these distinctive government models, as in Sakai (69).

We shall mention some detailed examples of the transfer of specific techniques in the cotton spinning and weaving industry from Britain to Japan. In 1889, all modern cotton mills in Japan, more than twenty, introduced hook-up reelers for spinning by importing them from Britain. From 1889 to 1914, the number of modern mills increased by 450% exceeding 90 factories at World War I. However, only four hook-up reelers were imported from Britain during this period, while the rest were manufactured locally (67, 70).

We may divide the transferring hook-up reelers technology from Britain to Japan into three stages: first, the technology was imported in 1884 for five years, followed by the second stage of importation plus local manufacturing since 1889 for eight years. Finally, in 1897, Japan depended totally on the domestic manufacture of this technology. Not a single piece of this technology was imported in more than ten years. In other words, the technology was fully adapted in Japan nearly 13 years after entering the country. By adaptation, we do not mean introducing this technology to the country, but rather its manufacture within the country without dependence on foreigners and imports.

Regarding the technology transfer of the bundling pressure technique transfer in the spinning industry, 86 factories in Japan imported this technique from Britain in 1906. However, only one bundling pressure device was imported during the next nine years, while the local needs were covered by local industry (67, 70).

We may analyze the stages of transferring the bundling pressure technology from Britain to Japan (though the separation between the stages was less precise than the previous technology). The import stage began in 1884 for seven years to meet the domestic needs. In 1892, the second phase began, which was local industrialization with import, so one-third of the domestic needs were fed locally, and two-thirds were imported. This stage continued until 1906 and for 15 years. Then Japan reached the third stage, which depended on the total domestic production of this technology in Japan (and dispensed with import), starting in 1907. Japan recorded only one import application during the following ten years. In other words, in 45 years, Japan owned a technology that did not exist before and moved from total dependence on importing to local manufacturing and exporting cotton machinery. For example, in 1909, Japan had 1,800,000 spindles, of which 1,570,000 were imported from one British company, the Platt Brothers of Oldham (67, 70).

~≈~

THE ROLE OF THE JAPANESE GOVERNMENT IN THE COTTON INDUSTRY DURING THE TOKUGAWA AND MEIJI ERAS

The active contribution of the Japanese government since the era of Tokugawa, Meiji, and his successors, part of which is explained in this chapter, helped transfer the technology (imitation and copycatting). The Boren Cotton Industry Association also disseminated knowledge and data about the cotton industry and an overview of the latest machinery and its suppliers or local manufacturers. This Association also ranked companies according to their efficiency in extracting textiles from cotton. Because the cotton used was imported, it was a practical step to encourage saving industry inputs.

One of the measures that helped in the renaissance of the cotton industry in Japan was the establishment of the Engineering Council, a college to teach the skills of the textile industry and how to operate and repair the spinning and weaving machines imported from the West. This college then turned into the first university in Japan named Tokyo Imperial University in 1886, today's Tokyo University. Today, it is one of the most prominent universities in Japan and East Asia. In addition to establishing a company specializing in importing cotton suitable for steam cotton spinning and weaving machines, Japan Cotton Trading Co. specialized in its early days in importing cotton from the USA (70).

At the end of the Tokugawa era, there were 3,000,000 Japanese employed in agriculture and the cotton industry, making up 10% of the population. The value of the output of cotton mills in Japan increased about 40 times in 30 years, from 32,000,000 yen in 1895 to 1,200,000,000 yen in 1925. Thanks to the progress in the cotton industry in Japan, historians called Japan at that period "Britain of the East" and Osaka, the city in which the cotton industry flourished, "Far East Manchester" (70).

Kagoshima Mill was the first modern spinning mill in Japan established by Shimazu Nariakira, chief of the Satsuma clan, on Kyushu Island in Kagoshima City and employed 40 workers there. Although it was established in 1850, it adopted traditional methods at the time. Then in 1865, modern machines were imported from Britain from Platt Brothers in Lancaster, Britain, at a cost of £10,000. The machines were operated thanks to the support of five British engineers in 1876. The mill had 20,000 Japanese workers and 3,648 spindles. Even though it was the first and largest in Japan at the time, it was very small in European standards, especially in Britain. However, this mill failed for several reasons, including the overemployment of more than twice the required number of workers. The tribal decision-making mechanism also affected its performance. Add to these its distance from the main markets for selling cotton in Japan, which raised the cost, and from the ports designated for importing raw cotton suitable for modern machines. Thus, the local leaders of the Satsuma clan set up a second modern mill closer to the market. The Sakai factory, established in 1871, was the second cotton mill in Japan with 4,000 spindles, but it was not also profitable until it was sold to the private sector in 1878 (70).

The third mill established in Tokyo was Kashima in 1874, but it was tiny with 576 spindles and fifty-two workers. Until the end of the 1870s, all these mills, with a total capacity of about 8,000 spindles, were unprofitable because the cotton textiles imported from Britain arrived at a lower cost to the final consumer in Japan's markets (70).

Then, during the Meiji era, the government supported ten small factories throughout the country with a capacity of 2,000 spindles. After the success of the Osaka Company, in just three years (1885-1887), nine mills with a capacity of 10,000 spindles and higher were established, all of which used the Osaka Mill as a successful model to imitate. Before the Osaka Company, there were seven mills in Japan in 1881 with a capacity of only modern 16,000 spindles. In 1889, the number became 28 with a capacity of 215,000 spindles, an increase by six times. Then the number grew to 1,200,000 spindles in 1899, with 83 mills. In 1890, a cooperation took place between three mills (Amaga-saki, Settsu, and Hiraro) with a total capacity of 42,000 spindles under the leadership of engineer Kikuchi, who studied for six years in Britain at the 1880s. This cooperation ended with the merging of the three companies into the Greater Japan Spin-ning Co. led by Kikuchi in 1918 (62, 70).

Nakaoka *et al.* pointed out in 1988 that Japan was able, after importing modern weaving and spinning machines, to innovate and develop these machines to increase production capacity, thus gaining comparative advantage in exporting later to the countries of the world. To compare the economic performance of Japan's textile industry returns at the province and region level, Kyoto can be used as a model. In 1883, the textile indus-try's revenues constituted 73% of Japan's industrial economic returns, at 6.3 out of 8.5 million yen (71).

CHAPTER FIVE

INTELLECTUAL PROPERTY LEGISLATION AND AWARDS IN JAPAN SINCE THE NINETEENTH CENTURY

INTELLECTUAL PROPERTY LEGISLATION IN JAPAN

At the outset of the reign of Emperor Meiji in 1869, the Western influence was considerable on all levels of life in Japan, whether political, social, organizational, or cultural. Emperor Meiji made the West a model in almost every industrial and social activity. However, not all of such influence had a positive impact on the one hand, and it was not as per the will of the Japanese people or the Emperor. Instead, the influence was imposed on the Japanese by force, the occupation of ports, and military bases for foreign powers. At that time, Western powers were grappling for influence over Japan, China, and the rest of East Asian countries (72-74).

Emperor Meiji received unlimited support for implementing the Western policy in Japan and could not free himself from the influence of Western powers. His foreign advisors dictated policies up to the outbreak of World War I. However, the plurality of influence was in the emperor's favor. All the Western powers wanted him to be on their side, a positive loophole he and his successor exploited to pass policies in favor of Japan and support the most beneficial Western influence to the country. The policy for nearly four decades was to balance influence to obtain the most significant benefit for the nation (72-74).

Japan was the first country in the East to implement intellectual property policies and set regulations for monopolizing industries as per the European and US model. In 1871, the Provisional Regulations for Monopoly was issued only to be suspended a year later for lack of applicability. Its amendment was postponed until the issuance of the Patent Monopoly Law in 1885, the Trademark Law in 1884, the Designs Law in 1888, and the Utility Model Law in 1905 (75).

In 1899, Japan joined the Paris Convention for the Protection of Industrial Property (issued in 1883), which allowed foreigners to benefit from the local intellectual property system of the States Parties to the Convention. These laws and regulations were just the nuclei of current regulations that were revised, refined, and reissued. For example, the Patent Monopoly Law was revised several times, including in 1921, during the Taisho era, and in 1959, during the Showa era. (75).

Administrative bodies were also established to support the innovation system, including the Patent Bureau in 1933 with 600 employees. Takahashi, the Director of the Patent Bureau, was reported to have said, "*We looked around at what made nations great so that we could become like them....and we found that the reason was patenting, and we will get pate*nts" (76, 77).

The analysis of these regulations and laws would show that their essence, since their inception, was what was stated in the government plans in Japan, i.e., the local enrichment of industry and defense. The government had two priorities: to build the country's defense system and provide it with the latest technologies produced locally and substitution of imports by local products to support the economy, i.e., the *kokusanka* policy, or self-reliance in arms production (78, 79).

The substitution of imports by the local product was a European concern from the fifteenth to the nineteenth century and the USA upon its independence. Japan followed the path of imitation and copycatting of the thriving Western industries in the nineteenth century. Europe did the same during the previous centuries, but by imitating the flourishing industries in the East. It seems that history repeats itself, and the success mechanisms remain the same over time, i.e., to follow the successful model by imitating it, and thus countries progress and compete. Excellence remains in the creative content offered by nations, in addition to their speed in simulating and copycatting others' successful experiences.

An article in Nature in 1935 pointed out that all patents in Japan for 20 years after the issuance of the Patent Monopoly Law in 1885 were merely imitations and copies of successful Western techniques (77).

✥

ANALYSIS OF INNOVATION OF EARLY PATENTS GRANTED IN JAPAN

After issuing the Patent Monopoly Law, samples of the first patents granted in Japan would identify the amount of copycatting and imitation. On August 14, 1885, the Japanese Patent Bureau granted the first seven patents in Japan. Patent No. 1, granted to the Japanese Zuisho Hotta, was a corrosion-resistant coating for iron ships. This mixture comprised hard metal salts like iron, lead, gum, and thymol. He registered it in the USPTO

25 years later and obtained patent No. 916869 for the same innovation in 1909, entitled "Anti Fouling and Anticorrective Composition" (80, 81).

The innovation analysis of this patent would show that a mixture of metals with a dye to become resistant to corrosion was invented by William Beale in 1625, more than 250 years before Hotta. Beale mixed iron and copper powder with cement to make the pigment more corrosion resistant. In 1870, more than 300 mixtures were added to dyes to make them corrosion-resistant (82, 83). This proves that the first innovation granted in Japan was not for inventiveness but for imitation and copycatting of what existed outside Japan or arrived in Japan as a product. Still, the way of its production was unknown to the Japanese. The USPTO granted him the patent to reveal the secret of the mixture used in Japan.

Three of the first seven patents, 2, 3, and 4, were granted to Takabayashi Kenzo for tea production machines. Developed copies of these machines were produced later in 1898, and a sample of them is still on display in the Kawagoe City Museum. These machines sorted and expanded tea leaves after harvest, facilitating the sorting process and saving large labor numbers (84, 85). However, these machines were mass-produced in India, and many of their designs were British patents when India was a British colony. These machines did not exist in Japan and the

Japanese did not see them despite their production of tea, which made the tea production operations in India prosper a lot compared to Japan. This prompted Kenzo, and others later, to try to imitate this machine. In other words, these machines were new to Japan, but they were already innovated and had wide commercial uses in India and Britain.

One of the innovations that brought great material benefits in Japan with the Patent Monopoly Law was a technology for making salt using wind energy, which was granted to Tomogorō Ono (1817-1898). Solar radiation in Japan was low due to the abundance of clouds and winds, which made the drying of seawater to extract salt very expensive and time-consuming. Ono borrowed a German innovation of the sixteenth century before the existence of the patent law in Germany, where it dried seawater vertically instead of the usual horizontal methods by placing it in several basins on top of each other and exposing them to the wind, so the drying process was done using wind energy instead of solar radiation. Ono's invention was nothing but copies of the German invention that existed more than two hundred years ago (76).

This shows that the first innovations in Japan were nothing but imitations of innovations that had already existed.

PATENT ATTORNEY LAWS AND REGULATIONS IN JAPAN SINCE THE NINETEENTH CENTURY

A set of government measures also helped grow the Japanese innovation system, including privatization of the government factories and mines at low prices in the 1880s. The private sector became a significant player in building the economy and replacing imports through local industry (73).

Japan also introduced the system of the limited liability companies, a format borrowed from Germany, which issued the first law of limited liability companies in 1892 (86). A group of systems, such as those of employment and monetary policies, were also issued to support the economy, let alone the direct laws, such as the Patent Attorney Regulations that began during the Meiji era. The Patent Attorney Registration Law was issued in 1899. It was refined during the Taisho era (1912-1926) (76).

Patent attorney laws existed in Western Europe and the USA, so Japan imported them, adding a local flavor. The most significant concern for inventors was to sell their patents after they were granted to the private and public sectors inside Japan. The attorneys followed up the procedures for granting patents and communicated with industry owners and relevant investors to sell these patents. Their role continued until after 100 years of issuing the Patent Monopoly Law, thus distinguishing the Japanese innovation system from its counterparts in the West. It contributed to raising the sale of patents from less than 5% before 1900 to more than 25% in 1924. More than 400 patents were sold in 1924, out of nearly 1,600 patents granted in the same year. The average cost of selling a patent was 5,500 yen per patent. The number of registered patent attorneys increased from less than 20% per year before 1900 to more than 200% after 1925, a tenfold growth in 25 years (76, 87).

Patents increased from a few hundred (1885-1900) to more than 3,500 annually after 1925. To compare patent filing fees with patent attorney fees, patent fees relied on the protection time: 10 yen for five years, 15 yen for ten years, and 20 yen for 15 years. As for attorneys' fees, they varied, but 100 yen was an average at that time for each patent. Two decades after the issuance of the Patent Monopoly Law, the patent protection period was standardized to 15 years. Although patent fees were meager compared to Western European countries, such as Britain, France, and Germany, they were almost three times the fees in the USA at that time (76, 87).

⤞

CONTESTS, EXHIBITIONS, AND AWARDS TO ENCOURAGE INNOVATION IN THE NINETEENTH CENTURY JAPAN

One of the most prominent things that Japan made during that period and decisions supporting intellectual property, financial policies, and government programs to establish industries that replace imported products was the strategy of innovation exhibitions and competitions. Japan's policies in this regard were indeed unique; historians have not recorded any activity of this size and intensity in any other country (75).

Indeed, the idea of competitions and exhibitions for innovation and industry was not new, and the Meiji government did not suddenly introduce it to Japan. Fairs and competitions had existed in Japan, and governments during the Tokugawa era adopted many activities in this field. Western countries were also active with these activities, as shown below.

When two hundred English sailors died (considered as lost) in the fog in 1714 because the captain was unable to determine the latitudes, Britain announced a prize of £20,000 [a big prize at the time worth more than £10 million today] for the inventor of a tool or the machine that helps determine the latitudes west of the ocean. John Harrison, a watchmaker, invented a device that took seven years to build to determine latitudes at sea. However, the first model was impractical. He developed several models. His fourth model, developed 35 years after the first attempt, was a success, and he received the award after that in 1773. In this way, Britain used awards to encourage innovation in the eighteenth century (88, 89).

Another example in France was in the nineteenth century. When Louis Daguerre invented the portable three-column camera [the eponymous daguerreotype], the French government rewarded him in 1839 with a lifetime salary of 6,000 French Francs per year [Franc = 4.505 grams of silver] in exchange for making his innovation available to the public (90-92).

These are models of competitions and awards for innovation that preceded Japan's innovation period under analysis. These models may have inspired the Japanese leaders to adopt a competition strategy to support innovation. However, Japan entered this field powerfully and impressively, as the scale of the activities was unprecedented, as the details below would show.

The innovation competitions, awards, and exhibitions had two objectives. The first was to display the innovations and techniques to the masses to encourage them to emulate and develop them, in line with the famous saying, "a picture is worth a thousand words." Secondly, it was an excellent chance for marketing for innovators and their innovations. Awards in Japan were symbolic: shield, medal, badge, cup, paper (certificate), or a nominal cash prize. Given that the marketing costs of innovations were high, exhibitions and contests provided a low-cost marketing platform, especially for individual innovators.

Japan realized the technological shock (its sense of the wide technological gap between it and foreigners) after it participated in some exhibitions and competitions after opening up, especially at the beginning of the Meiji era, where innovators, mission staff, and representatives of the Japanese government in foreign exhibitions saw what fascinated them and proved to them their innovations were inferior. These included the exhibition in Vienna (Austria) in 1873, in Philadelphia (USA) in 1876, and in Paris (France) in 1878 (79).

To familiarize with the extent of what Japan did in this regard, we will highlight glimpses of the accomplishments during the 25 years from 1886 to 1911. Japan organized nearly 8,500 innovation competitions, nearly ten million innovation and industry fairs, and awarded 1.2 million prizes. With an average of 141 winners in each competition and 340 competitions per year, Japan held one competition per day over different regions, villages, and cities (78).

As for exhibitions, 396,000 exhibitions were held per year (an annual average during the period mentioned above), a daily average exceeding 1,000 exhibitions per day. These figures are enormous and indicate the seriousness of using this activity as a policy and strategy to encourage and support innovation. The turnout of these events was high. For example, there were 17 million entry tickets for these events during the 13 years from 1886 to 1898, with an average annual rate of one million and a quarter tickets. However, the fees collected for these entry tickets did not cover the costs of these events and were not initially designed for this purpose. Instead, they were an additional tributary to cover part of the events' costs. In addition, the average cost of setting up such an event was lower in Japan compared to France (50%), Britain (64%), and Germany (90%) (78).

Nicholas (2013) studied the relationship between the number of innovation competitions in Japan and the number of patents filed during the ten years from 1885 to 1908. He attributed this connection to integrating innovation systems to support inventors and implement activities that encourage innovation. Moreover, the connection between the increasing companies and patents granted in Japan is noticeable, as both increased from 1885 to 1910 by about 6 to 7 times. In short, innovation increases the chances of establishing companies (79).

CHAPTER SIX

CHAPTER SIX

TECHNOLOGICAL ABSORPTION IN JAPAN AND ITS ECONOMIC IMPACT

Compared with many countries, Japan had been quick to copy and emulate the successful innovations of foreigners (93). We will provide an overview of Japan's performance in this field, but we will try to resolve the confusion in some common concepts and terms before we start.

The concept of technology transfer, diffusion, and adaptation is synonymous with the ability to copy and imitate foreign innovations, except for what is invented inside the country, such as moving from a research center or city to another or company to another within the same country.

A common confusion is also what Comin and Hobijn reported in their 2008 research claiming that transferring or adapting technology was the beginning of its entry or use within the country. However, this is not the true meaning of technology transfer and adaptation. It is the ability to reproduce technology inside the country after importing it or after it had reached the homeland in any way (such as stealing it, for example, when the Allies stole Freya radar from the Nazi forces or when the Byzantine emperor stole larvae silk from China) (94).

Importing technology is bringing it into the homeland. Unfortunately, many researchers and economic analysts err in describing the import of technology or its widespread use as technology transfer, a confusion we had to clarify here. Many technologies were imported and used in countries that could not reproduce them. Importing technology is not necessarily a step to be followed by local manufacturing of the same technology. It may just end up being used.

The following example should illustrate the difference between technology import and technology transfer. When a country imports a car, plane, and submarine. It introduces this new product to its territory. However, this does not mean that the technology of making the car, the plane, and the submarine is transferred once the product is imported, even if its use spreads within the country.

Accordingly, the concept of technology adaptation lag is the time required for a country to bridge the technological gap between it and the first foreign country to innovate the targeted technology. It is the time from the date of the first innovation of the technology and the date of its first copies (reproduction) within a country. However, it is not the time from the technology innovation and its entry into a country, as indicated by some of the above sources.

Although Comin and Hobijn's 2008 research is valuable and has set the time for the technology adaptation in 166 countries in 15 technological fields over 200 years, researchers have forgotten that the time of arrival of technology (or its spread after its arrival) to a country is not an indication of the ability of this country to remanufacture and reproduce this technology (94). Many researchers have fallen into this confusion, which made the scientific community miss valuable insights about technology adaptation in much research.

Furthermore, technology ownership or acquisition through purchasing or importing does not mean the ability to remanufacture or reproduce it locally. Accordingly, owning technology is not an achievement for any nation unless it can reproduce and remanufacture it, which is the first step in the path of innovation.

To study technology transfer, it is necessary to focus on a country's technology absorption capacity to know the reproduction speed of innovation in the recipient country. technological absorption consists of institutions, regulations, policies, and financing necessary to turn a foreign innovation into local production. In other words, it is the ability to bridge the technological gap of the foreign product which will become domestic/indigenous technology.

When a country reproduces and copies foreign technology, it can absorb technology, an ability that varies between countries and thus reduces or increases the time of technology adaptation or transfer. Japan possessed a high technological absorption capacity that enabled it to convert many foreign technologies into a domestic industry in short periods compared to the rest of the countries in the same period and geographical area (such as China). For example, Western technologies entered East Asia at the same period and contact with Japan was almost cut off between the seventeenth century to the middle of the nineteenth century. How could Japan transfer these technologies faster compared to the countries of East Asia, even though the rest of Asian countries remained open to the West (or cut off for shorter periods)? A later chapter will discuss Japan's technological progress over the rest of East Asian countries, or what is known as the 'flying geese' paradigm.

Although foreigners were allowed to file patents in the Japanese Patent Bureau, the Japanese increased their share of innovations at the beginning of the twentieth century before World War II in 1939 (76). It became clear that Japan moved from imitating and copying Western technologies after the Meiji era to technological advances in targeted foreign technologies. Japan's policies of copycatting and imitation yielded fruit after World War I; Japan achieved self-sufficiency in many technologies it was importing. It also began to invade global markets with its products.

Technological progress can be expressed in changes in the production factors, either by producing new products or introducing new processes for the pre-existing ones. Japan reached innovation after imitating and copying faster than Europe in the past and East Asian countries in the present.

Brown (2005) indicated that the time of technology adaptation in Japan in mining shrunk from more than 100 years for coal-burning furnaces technologies (160 years) to less than ten years for Stassano Electric arc furnace technology (73).

The acceleration in making metal steamships was also noted compared to wooden steamships. Remanufacturing operations began after importing directly from France. Thus, as we mentioned at the beginning of this chapter, Japan became one of the first countries to produce iron-armored steamships.

Technological progress in Japan was accompanied by an increase in labor productivity in the country's economic system.

The average worker's productivity rose from 155.7 in 1887 to 420 in 1915, nearly three times (73).

❧

"FLYING GEESE" ECONOMIC MODEL

The Japanese economist Kaname Akamatsu introduced the flying-geese model in the 1930s to explain the economic progress of East Asian countries. It interprets Japan's superiority, technological progress, and economic growth over East Asian countries. Japan was ahead of the countries of Asia in technological development and economic growth, followed by the countries of East Asia, which made Japan at the forefront of the flock of flying geese located at the head point the English letter "V" of the flock of geese, followed by the rest of the East Asian countries, such as Korea, China, Indonesia, Singapore, and Malaysia (95, 96).

In that period, East Asian countries were trying to emulate, imitate, and copy the techniques and policies of Japan, which enabled them to develop gradually. The more copies and imitations from an Asian country to the Japanese model, the closer it would be to technological progress and economic prosperity. The East Asian countries replaced the import of Western technologies with local industries, and then at an advanced stage, they began to export their local industries to Western countries.

This economic theory divides the stages of economic growth in latecomer countries: importing technology or innovation, local manufacturing and production, and export. It reinforces the concept presented in this book that imitation and copying are a stage of economic progress and a path to innovation later. The higher the copying, imitation, and replication of a superior nation, the faster economic prosperity and technological progress (95-97).

THE ECONOMIC IMPACT OF IMITATION AND IMITATION IN JAPAN IN THE TWENTIETH CENTURY

Due to technological progress in Japan, it increased its share of world exports from 2.9% in 1960 to 9.8% in 1985 when Japan's exports accounted for nearly a tenth of the world's merchandise exports, an indicator of productive strength. This high share of exports was partially taken from the share of many European countries (98). Porter and Sakakibara (2004), in an analysis of economic growth in Japan, indicated that the nature of competition in Japan was based on the high capacity for imitation and copycatting (98).

In the post-WWII period, it took Japan nearly 15 years to recover from the effects of the war. After that, it issued a series of decisions and launched a set of policies to develop the innovative local capacity; among the most prominent of these policies was a prohibition on importing 474 commodities and the entry of foreign capital in 1961 (98). These policies aimed to strengthen local production, imitation, and simulation and protect the local industry. However, such exceptional measures did not last long. After adding the innovative capacity to local products, prohibition was gradually lifted, and restrictions were eased over time.

CHAPTER SEVEN

CHAPTER SEVEN

~≈~

JAPAN'S INNOVATION AND TECHNOLOGICAL PROGRESS AFTER WORLD WAR II

JAPAN'S TECHNOLOGICAL PROGRESS IN MOTORCYCLE INDUSYTY

How did Japan enter the global competition in specific industries? How did Japan develop from imitation to innovation after World War II? We chose the following industries to illustrate Japan's shift in innovative capabilities, namely motorcycles, soft drinks, video games, robotics, and semiconductors and computers.

In the early 1940s, motorcycles were a relatively simple technology, consisting of a bicycle containing a small engine with less than ¼ horsepower (hp). It spread in Europe for its ease of manufacture and low cost to the end consumer, making it an alternative low-cost means of transportation for individuals. It was not manufactured in Japan on a large scale at the time. To highlight the history of the motorcycle industry and how this technology entered Japan, we need to go back to 1872.

The first motorcycle was invented by German Gottlieb Daimler and Wilhelm Maybach in 1872. They produced a single-cylinder bicycle engine and patented it in 1885. They received a series of innovations from the German Patent and Trade Mark Office in the bicycle industry for the following patents: DRP68492, DRP70577, and DRP75069 (99-101).

In 1909, the Japanese Narazo Shimazu designed and manufactured the first local motorcycle in Japan after establishing the Nihon Motorcycle Company, using a model of bikes imported from Germany. A model motorcycle of the Hildebrand & Wolfmüller, the first motorcycle made in Germany, was displayed in Japan in 1896. Jumonji Nobosaka, one of the owners of Jumonji Trading Co, imported it. Shimazu used the German model to remake a motorcycle (102, 103).

In the mid-1940s, Britain (1946) was the largest manufacturer and producer of motorcycles globally, followed by France, Germany, and Italy. In Japan, in 1946, the annual production did not exceed 200 motorcycles of a low-quality (primitive) type for local use only. (104).

Three years later, in 1950, 27 motorcycle manufacturers in Japan produced 9,000 units per year. This number rose to 127 companies in 1952. The local production rose over time, up to 1.5 million in 1960 when Japan became the world's largest exporter of motorcycles surpassing Britain, France, Germany, and Italy (the top four motorcycle-producing countries at the time) (104).

However, the quality of Japanese motorcycles was low, and they were known as fakes; they were unexportable. In 1960, the export percentage did not exceed only 4%. However, the technological content and internal development gradually rose, allowing Japanese motorcycles to compete globally. In 1965, the exported local production reached 45% (104).

From 1955 to 1965, Japanese companies shifted from importing parts and imitating designs from Europe to producing parts locally and modifying designs with a Japanese flavor. The Japanese motorcycles gained higher acceptance among foreign consumers, thus increasing the demand outside Japan (104).

One factor contributing to this technological development was some Japanese companies specializing in motorcycles, such as Honda Motorcycle Co., which produced its unique Super Cub C100 model, 4.3 hp, in 1958 (105). This innovative

Japanese model encouraged the rest of the local companies to emulate and copy the successful model, which promoted and developed the local industry due to local competition. Honda sued its competitor Yamaha for copying this innovative model, prompting competitors to make changes and develop the Honda model to compete. Patent rights and State protection prompted the big three competitors, Yamaha, Suzuki, and Kawasaki, to develop their innovative models to stay in the market. For example, Suzuki Corporation introduced a group of developed models, including the T20, in 1965, which later invaded global markets (104, 106).

This, in turn, led to the entry of a developed group of motorcycles to the local market first in the early sixties, and then they found great acceptance in the global markets. Japan's annual production of motorcycles rose to nearly 3 million by the early 1980s.

This overview of Japan's capabilities in motorcycles shows how Japan began in 1896 importing a technology it did not own and then imitating it in 1909 while importing spare parts and assembling them in Japan, and then the advanced stage of intensive production, manufacturing, and export in 1950-1960. It moved to innovation in the early 1960s for more than twenty years as a final fourth stage. This shows Japan's rapid technological adaptation ability, which has allowed it to dominate this industry. Japanese motorcycles are still highly accepted in global markets due to their high quality.

❧

TECHNOLOGIES FOR MAKING SOFT DRINKS IN JAPAN

Before analyzing the innovation in the soft drink industry in Japan, we will review the innovation in the world in this field to learn how it reached Japan first, and then how Japan emulated this industry and contributed to its development later.

In 1767, Joseph Priestley of Britain produced carbonated water (containing dissolved carbon gases) like naturally carbonated water extracted from some natural springs and wells. Before that, the world did not know how to produce water containing carbon gases (107, 108).

In 1771, the Swedish Torbern Bergman produced carbonated water similar to the English innovation by adding sulfur acid to extract carbonates from limestone and then adding it to water (109-111).

In 1783, Johann Schweppe established a company in Geneva, Switzerland, to produce carbonated water with the exact mechanism as earlier innovators in England and Sweden (112, 113).

Since it was common in medical science that carbonated water was of great health benefit, physicians prescribed it to patients and the elderly to improve digestion and relieve fatigue. In 1807, Dr. Philip Syng Physick in America added a flavor of fruit and herbal extracts to carbonated water to make it more palatable to patients and increase its consumption. However, this was an individual effort added in a pharmaceutical lab rather than a mass commercial production (114, 115).

Schweppes Corporation, founded by the Swiss Johann Schweppe, referred to above, added a small amount of ginger to carbonated water in 1851, and this invention was attributed to Thomas Cantrell (116). Accordingly, the first soft drink containing additives with water was produced in 1851 under the name Ginger Ale by Schweppes Company. This innovation inspired a swarm of companies to change the added ingredient and invent new soft drinks; innovation continues in this field.

In 1866, the American James Vernor created a relaxing soft drink containing ginger and vanilla called Vernors ginger ale. Then the Ecuadorean of Italian descent, Juan Fioravanti, invented his soft drink containing apple fruit extract added to carbonated water in 1878 and called it Fioravanti (117, 118).

In 1885, the Americans Augustin Thompson and Charles Alderton invented new soft drinks. The first produced the Moxie soft drink, which contained the roots of the Gentiana plant. The second created the famous drink, Dr. Pepper containing a mixture of flavors (more than twenty fruits) added to carbonated water (112, 119-121).

In 1886, American John Pemberton invented Coca-Cola by mixing Erythroxylum coca leaves with cola nuts. The cola nuts had a bitter flavor and a high caffeine content, hence the name cola. The name coca was also derived from coca leaves, which contained a percentage of cocaine (9 mg per bottle). Coca-Cola has been one of the most famous soft drinks in the world. It is reported that John Pemberton was one of the soldiers returning from the American Civil War (1861-1865) between the North and the South. He suffered several severe wounds and was treated with morphine to calm the pain after the defeat. He found an imported French product containing French wine and cocaine. He remanufactured it for individual consumption as a substitute for morphine and sold it to others until he was prevented from doing so (122-124).

He produced an alcohol-free product, Coca-Cola, and cocaine was not banned at the time, so the company continued to produce and sell it for 18 years, but government pressure began, followed by the official ban in 1914 with the issuance of the Harrison Narcotics Tax Act that regulated and taxed the production, importation, and distribution of opiates and coca products. Therefore, the company produced, since 1904, the Coca-Cola product, free of cocaine, by exposing the coca leaves to high heat to get rid of the cocaine extract (122-124).

In 1893, the American pharmacist Caleb Bradham invented a mixture of water, sugar, caramel, and lemon oil with cola seeds and walnuts and called it Brand's Drink. He then replaced it with Pepsi-Cola to emulate the success of Coca-Cola (125-127).

In 1907, the Canadian John McLaughlin obtained a patent to produce carbonated water containing ginger and sugar called Canada Dry Gingerale. About 17 years before (in 1890), he owned a company to produce additive-free carbonated water (116).

In 1929, Charles Lipper invented his product containing lemonade, sugar, and carbonated water, known today as 7up. In the 1940s, Barney and Ally Hartman created a 7up containing lemonade called Mountain Dew (128).

This historical view tells us about the history of the soft drink before it entered Japan. In 1884, a Scottish pharmacist named Alexander Cameron Sim brought a soft drink containing lemon extract to the Kobe foreign settlement east of the Port of Kobe, Japan. It is believed that this drink was the first flavored soft drink that entered Japan and was later known as Ramune. This product is still prevalent in Japan today (129).

The soft drink industry began in Japan. When a market-successful product appeared, many companies were created to simulate it to gain market share and meet demand. Thus, the soft drink industry in Japan developed up to innovation and advanced technological progress. Let us compare several new soft drink products that entered the market annually in Japan during the two decades 1980-2000. The annual average amounted to nearly a thousand new products, which exceeded the USA, the most prominent soft drink market today. The new soft drinks in the USA did not exceed 700 per year (129).

Japan introduced many new products to the soft drinks market, most notably the so-called honey drink, carbonated coffee drink, Japanese tea carbonated drink, black tea carbonated drink, and new-flavored carbonated water. The internal competition between soft drink companies was intense. If we compare imitation in the Japanese carbonated products market regarding what was known as the honey drink, Nisshia Seiyu Company introduced the first drink of this kind, called Hachimitsu Dori, in 1985. Suntory Company introduced a similar product in 1986 called Hachimitsu Lemon, then 28 companies appeared before the end of 1989 offering similar drinks (129).

This fierce domestic competition and copycatting the products of successful competitors was the reason for developing markets and constantly innovating new products for the innovator to remain in the lead in the markets. What happened in Japan has been taking place between the two US soft drink giants (Pepsi-Cola and Coca-Cola) that imitate each other's successful products, albeit with a slight change, to evade flagrant imitation (130, 131).

The global soft drinks market reached nearly US$ one trillion in 2020, and Japan was among the top four countries globally (along with the USA, China, and Canada) in market share. Back to the underlying concept, Japan did not invent this industry but managed through imitation, copycatting, and simulation to excel economically over the countries that preceded it in inventing this technology (132).

⟿

INNOVATION AND IMITATION IN THE AUTOMOTIVE INDUSTRY IN JAPAN

Although historical sources do not identify the first car (whether with steam, gasoline, or diesel engine) imported to Japan, nor the country of manufacture, estimates indicate that the import of the first cars to Japan began after 1853 during the Tokugawa era. During that period, the Shōgun was interested in public transportation. In 1866, he laid out a plan to construct the first railway between Tokyo and Yokohama (133, 134). It is believed that the first cars entered then with representatives of foreign countries as gifts to the local rulers and clan chiefs.

In 1904, Torao Yamaha manufactured the first Japanese steam engine car, a small bus carrying ten passengers. In 1902, Komanosuke Uchiyama imported a gasoline engine from the USA to install it in a locally-made car. In 1907, as an engineer in Tokyo Motor Vehicle Works, he completed the first Japanese-made gasoline engine automobile, the Takuri, with gasoline engines imported from the USA (135-137).

In 1910, the Kunisue Automobile Company developed the Tokyo car (138). In 1911, Masujiro Hashimoto, a mechanical engineer in the Japanese Ministry of Agriculture and Commerce, was sent to the USA, worked in steam engine factories in New York, and established the Kaishinsha Motorcar Company, which produced the DAT-31 car in 1914 then the DAT-4 in 1916. After repeated mergers, the company's name changed several times until it finally became Nissan (139-141) in 1933.

In 1917, the founder of Mitsubishi produced the company's first car, Model A, with a speed of 100 km/h. Its production took three years (142, 143). In 1921, Junya Toyokawa, the founder of Hakuyosha Company, produced two prototypes of the Ales, followed by its successful model in the local market called Otomo in 1927. In 1936, the Toyota Industries Corporation began manufacturing its Toyota Model AA (144, 145).

By the end of 1935, Japan had 16 local car manufacturers annually producing about 500 cars from 1930-1945, compared to the import of the US cars to Japan, which exceeded 20,000 cars annually. Accordingly, the local production did not constitute more than 5% of the domestic consumption. From 1902 to before World War II in 1939, all cars manufactured locally in Japan for Japanese companies were still replicas of European and US products and did not live up to the taste of the Japanese elite. They had no market potential outside Japan. For example, in 1950, Japanese car exports did not exceed 3.1% of total domestic production (146-149).

The foreign companies (US and European) established car assembly plants in Japan to meet the high Japanese domestic demand, avert the high cost of sea freight, and reduce shipping time. In 1925, Ford established a factory in Yokohama, and its competitor GM followed suit in 1927 in Osaka, then Chrysler. For comparison, the production of US cars in 1924-1936 totaled approximately 209,000 cars, compared to 12,000 Japanese cars in the Japanese market. Japanese auto companies would not have competed without government intervention, subsidies, and decisions to find domestic alternatives in the early 1930s. Japanese companies were imitators in the auto industry (146-149).

Japanese car designs and models relied on Western models and designs, either through cooperation or simulation, imitation, and copycatting. Mitsubishi simulated the Italian company Fiat; its Model A was a licensed simulation of Fiat Tipo 3. Toyota imitated Chrysler's AA model under another license; Isuzu partnered with Britain's Wolseley Motors; Nissan imitated the British company Austin; while Ohta Jidosha imitated the Ford models (150).

Thus, it is clear that the Japanese auto industry could not innovate in its first 50 years but only provided replicas of engines, bodyworks, and designs of imported cars. However, in the mid-fifties and early sixties, domestic Japanese companies introduced innovations and remarkable car development.

In 1955, Suzuki introduced its 350-cc Suzulite, immediately succeeding in the local market. Then competitors followed by using the same engine power, as Fuji Corporation introduced the Subaru in 1955. Mitsubishi Corporation introduced its 500-cc Mitsubishi 500 in 1955. In 1966, Nissan introduced 1-liter Sunny, followed by Toyota's 1-liter Corolla (151-153).

In the sixties, Japan surpassed Germany in the number of cars produced annually. This development continued in the eighties until the total number of cars produced by Japan exceeded those produced in the USA. The peak of Japanese car production was 13 million. In the 1980s and 1990s, Japan had the highest share of car exports worldwide. Japan's production capacity was not only in number, but the quality rose to the level of luxury. Japan began to compete in this field in the 1990s after introducing brands, such as Lexus, Acura, and Infinity, luxury cars that rivaled the luxurious German cars Mercedes and BMW. In 2012, Japan's share of the global auto market was nearly a third (30%), and six Japanese companies were among the world's top 10 automakers (146, 154, 155).

An analysis of Japan's position in car innovation would show that it spent nearly half a century of imitation and copycatting to reach innovation. When it started innovation in the early 1960s, it took 20-30 years to have the top global market share in car sales.

This success in innovation in Japan in the automotive industry led to the bankruptcy of the US city of Detroit, home to the headquarters of the three major car companies (Chrysler, Ford, and GM). Unemployment rose to 30% in the city, and Chrysler filed for bankruptcy and was bought out by Italy's Fiat. The success of the industry of one country destroyed the industry of other countries (156).

❦

INNOVATION AND IMITATION IN JAPAN'S SEMICONDUCTOR INDUSTRY

The second technological shock that Japan had was in World War II. The Japanese were convinced in 1945 that their defeat was due to the opponents' superiority in technologies and sciences. This shock prompted the Japanese to imitate and copycat at a higher rate after World War II, and the US model was the model they adopted (157). European models were more attractive for imitation and copycatting after receiving the first technological shock in 1853.

Accordingly, transistors and semiconductors have been a target of imitation since their invention in the USA. This made the technological competition fierce in the 1950s for the transistor industry and fiercer in the 1960s over closed circuits until Japan overtook it in the mid-1980s.

Although the transistor, the basic unit in the semiconductor and electronics industry, was not invented in Japan but in Bell Labs, the scientific arm of the US AT&T telephone company between 1947-1948, Japan became the largest transistor-producing country worldwide between 1959-1960, thus excelling the country that invented it in production volume.

With the development of the semiconductor industry, which contains many transistors, Japan gained the highest global market share between 1993-1986. Also, in 1989-1990, it had more than 50% of the market in this industry. The rest of the world shared the remaining market share of less than 50% (158).

To get acquainted with Japan's success, we start with the history of this innovation in the USA, followed by the chronology of technology development in Japan until it surpassed the USA. In 1947, John Bardeen, Walter Brattain, and William Shockley created the first primitive point-contact transistor at Bell Labs. Shockley introduced the improved bipolar junction transistor in 1948, which entered production in the early 1950s and led to the first widespread use of transistors. The three later received the Nobel Prize in Physics in 1956 because these works paved the way for the invention of the integrated circuit (IC), later used in all electronic devices, such as phones, radios, televisions, computers, video games, etc. In 1949, German Werner Jacobi from Siemens obtained a patent for an integrated circuit containing five transistors. The American Sidney Darlington obtained a patent for installing a three-transistor chip in 1952, and Bernard Oliver also obtained a patent for placing three transistors on a semiconductor chip in the same year. Harwick Johnson received a patent in 1953 for a method of forming various electronic components-transistors, resistors, lumped and distributed capacitances- on a single chip (159-161).

In 1957, the Japanese Yasuo Tarui, at the Electrotechnical Laboratory of the Ministry of International Trade and Industry (MITI) near Tokyo, fabricated a "quadrupole" transistor, a form of unipolar (field-effect transistor) and a bipolar junction transistor on the same chip. In other words, Japan innovated in this field after nearly ten years of imitation, which began in the same year the transistor was invented (162, 163).

New knowledge rarely comes suddenly in a vacuum, but it develops through learning the existing knowledge first and then trying to add to it, develop it, correct it, and build on it. Imitation and copycatting of the existing sciences, not bypassing and ignoring them as if they had not existed, are the first steps to creating new science. The Japanese government significantly encouraged this industry by taking conducive measures, including banning the import of transistors and semiconductors. These included financial support, soft loans, and negotiations with foreign companies and governments to facilitate technology licensing to Japanese companies through its arm (Ministry of Agriculture and Commerce). Western Electric Inc. licensed the technology to Japanese companies in the early 1950s (164).

Sony, a startup at that time, had begun its activities by imitating and simulating the primitive tape recorder invented by AEG Co. in Germany in 1936. Sony's founders obtained an old copy that the Japanese army had used during World War II, and the company produced its first replica in 1949-1950. Later, Sony Corporation introduced the transistor into the radio industry instead of the old technology. It also developed the transistor for the television industry in the 1960s. In 1979, it produced Walkman TPS-L2, the world's first portable music player. Then in 1980, it developed the compact disc (CD) (165).

In 1956, Fujitsu established a transistor factory in Kawasaki to manufacture silicon transistors and Fukushima Factory in 1967 to produce audio devices using transistors. In 1980, Fujio Masaka invented the flash memory for Toshiba Company, which was the first tool for rewriting and saving information without turning off the device (166, 167).

The transistor innovation did not initially originate in Japan. Still, Japan's procedures for three decades made it excel in its industry. Whenever a new electronic product appeared, Japan would quickly copy and imitate it and excel in its production. The superiority came in two stages, the first with production intensity (in number) and the second, always later, high quality accompanied by some development and innovation in production methods. It is impossible to enumerate all of Japan's innovations and developments in transistors, semiconductors, and integrated circuits. However, we note Japan's high capacity of technological adaptation, which was later a cause for innovation accompanied by economic superiority and reflected in market shares referred to briefly at the outset of this chapter.

Before concluding this chapter, we cannot fail to refer to the Japanese government's effort to finance a vast research project for cooperation between Japan's five major electronics companies, namely Nippon, Toshiba, Mitsubishi Electric, Hitachi, and Fujitsu, with a budget of US$ 280 million for four years (1976-1980). The objective was to push these companies to stay ahead and add innovative content to the industry, especially in the very-large-scale integrated semiconductors (VLSI) used in computers and electronic products. This collaborative project increased the number of patents and domestic import alternatives. The economic fruit was evident since the mid-1980s (168).

ROBOTS AND INNOVATIONS IN JAPAN

Japan managed to capture 70% of the world's robot market by the early 1980s and produce three times the number of robots produced by the US in 1982. The number of robots in 1982 in Japan reached about 22,000, compared to about 6,800 in the US. Although the robot is a US innovation patented in 1961, Japan imitated this technology and surpassed the US in production in less than 20 years. Then Japan began to raise innovative content in robots, thus becoming a world leader in this field to the present day (169). The 2016 Market Report on Japan's Smart Robotics Industry indicated that Japan had the highest number of industrial robots of 295,000 in the world in 2014 (170). We will highlight the history of the technology invention, including its literary background that preceded the innovation, and then how it reached Japan until it gained industrial supremacy in this field.

The idea of robots began with a series of science fiction writings that preceded innovation and inspired scientists to produce robots and make them a reality on the ground, helping later generations to imagine their existence in the future.

The British Mary Shelley penned her famous novel, Frankenstein, in 1818, about a non-human creature made by a scientist through a scientific experiment (171). The writer may have been inspired by some of the ancient myths of the Greeks, such as Galeta, who transformed from an ivory statue into a human being, or the Hebrew myth of Golem, an animated anthropomorphic being in Jewish folklore, which is entirely created from inanimate matter (usually clay or mud). Shelley's literary work inspired other writers to refine the idea (172, 173).

The English writer Edward Ellis published a novel entitled *The Steam Man of the Parries* in 1868. This novel shows that the writer was influenced by steam machines in everyday use, which led to the laying off of many workers after the Industrial Revolution. Ellis developed an iron steam engine in the form of a talking man pulling carts instead of horses (174).

In 1907-1908, Lyman Frank Baum wrote *Ozma of Oz* and *Tik-Tok of Oz,* including an iron woodcutter, a straw man, and a man in the shape of a globe carrying a gun. All these characters walked and talked like human beings (175, 176).

In 1920, Karel Capek, the Soviet writer from Czechoslovakia, introduced the word 'robot' in *Rossum's Universal Robots* derived from 'service' in the Czech language. The Russian writer Isaac Asimov reinforced the concept of robots in his 1942 novel, *Three Laws of Robo*tics (177, 178).

These literary works honed and refined the concept of using a machine to perform various human roles and paved the way for the US investor Joseph Engelberger to adopt the first innovation in robotics (179). In 1954, George Devol filed a patent in the USPTO and gained patent No. 2988237 in 1961. With Joseph's support, George produced the first model of a robot (a robot arm) in 1959 for the product Unimate 01, which cost nearly US$5 million to produce the first model. Then, in 1961, he produced the Unimate 1900 Series, which Unimation Inc. sold to GM in 1962 (169, 180, 181).

Then a competing company that appeared in 1960, American Machinery & Foundry (AMF), produced the Versatran robot, marketed and installed in the Ford Motor Company production lines. In 1967, AMF presented a film about Versatran to Toyota, which produced its replica in 1968 (169, 182).

At the same time, Unimation licensed the robot's technology to the Japanese company Kawasaki Heavy Industries, and the company sold it to Nissan. Later, Mitsubishi and Fujitsu used the Unimation model to produce their first models through imitation (169, 182).

Japan quickly joined this field, as government helped speed up steps and reduce time, including the establishment of the Industrial Robot Round Table Corporation in 1971, which later turned into the world's first association for robots named Japan Industrial Robots Association under the umbrella of the Ministry of Trade, Economy, and Industry (METI). The association issued magazines, held conferences, distributed questionnaires, and statistics, and translated relevant foreign content to readers and businessmen of the industry. The Japanese government also provided loans to 13 start-ups and passed a law stipulating that the State would incur the costs of depreciation in robotics factories. It also provided soft loans to companies to purchase robots. By 1983, Japan had 150 robot companies, compared to 6 companies in the USA. The profits of major Japanese companies amounted to more than US$15 billion annually, approximately 15 times the annual profits of one of the largest US robotics companies, Cincinnati Milacron, which alone had 30% of the US market. Japanese robotics companies specialized in specific areas. By 1980, most of the production of robots was used in car factories. Nissan and Toyota used robots for labor-hazardous welding and Honda for painting and coating (169, 182).

It all started with science fiction and then technological innovation outside Japan before 1960. However, Japan started imitating and simulating these technologies quickly until it began to add innovative content to robots. Later in 2004, the Motoman Company introduced the NX100 robot, which contained 38 movement axes. In 2010, Fanue produced a robot that learns from surrounding vibrations and adjusts its speed (182).

In 1973, Waseda University produced WABOT-1, the first android/humanoid robot developed globally, which took six years to produce the initial model. It consisted of a limb-control system, vision, and conversation system. (183, 184).

The same university also produced its developed product, WABOT-2, in 1984, which played a musical instrument and was defined as a "specialist robot" rather than a universal robot-like WABOT-1. In 1985, in cooperation with Hitachi Corporation, the university produced the WHL-11 robot, a biped robot capable of static walking on a flat surface at 13 seconds per step; it could also turn (183).

In 1995 and 1997, Hadaly and Wabian robots, which could interact with humans, were produced. In 1996, Honda produced P2 and P3 robots that simulated humans and had better performance and speed of movement. Later, Osaka University produced its Actroid in 2003. It developed the humanoid robot DER1 and DER2 in 2006, then Actroid-F in 2010 (183, 184).

The conclusion in this section is how Japan led and excelled in robotics through a technology it did not invent, but through copycatting, imitation, and simulation up to the establishment of a local, national industry followed by a series of developments and progress to which Japan contributed before reaching technological excellence and the highest market share.

❧

IMITATION AND INNOVATION IN COMPUTER AND VIDEO GAMES IN JAPAN

Similarly, Japan did not invent video games but obtained the largest market share later, starting in the eighties. For comparison, the sales of Sony and Nintendo (two Japanese companies) exceeded the sales of Microsoft and Atari from 1970-2020. Nintendo made nearly five times the revenue of Microsoft for its Xbox product, the Xbox, at US$754 million, compared to US$149 million for Microsoft (185). In this section of the book, we will trace the innovation of video games in history, how Japan copied this innovation, entered this field, and then obtained the highest market share of sales later.

In 1958, William Higinbotham, a researcher at Brookhaven National Lab, created the first interactive tennis video game for two players (186, 187).

This game would not have appeared without the efforts of the German scientist Ferdinand Braun, who won the Nobel Prize in Physics in 1909 for developing the first model of the oscilloscope in 1897, developed later by others in 1931. This upgraded version was used at Brookhaven National Lab in the US to produce the first interactive game (187-190).

Another researcher at MIT Laboratories, Steve Russel, produced the Star Wars game Space war in 1962. These early works were in the laboratory (191, 192).

In 1972, Atari and its founder, Nolan Bushnell, produced the first video game outside laboratories in large, heavy-weight rectangular boxes called Pong. This first game was a 2D tennis game followed by arcade-type mobile video games placed in theaters and amusement arcades (193, 194).

German-American Ralph Baer produced Odyssey in the same year, the first home video game connected to the TV. He applied in 1971 and received patent No. 3728480 by the USPTO entitled "Television Gaming & Training Apparatus" in 1973. The Magnavox television company produced 350,000 copies of its first video game console (sales price of US$99) (195,196). The Atari company produced a video console for home games later, the Atari 2600 model, in 1977, which sold nearly 30 million copies (194).

Video games began to flourish in the early 1970s in the USA. The first video game consoles were imported to Japan in 1972, then domestic manufacturing operations began in 1973, followed by the first export of video games in 1974. In just two years, Japan manufactured and exported video games, but they were of low quality initially, and there was no demand for them outside the country. They were mainly used for domestic consumption (197-199).

In 1978 (six years after importing the first video game to Japan), Japan swept the world market and introduced its innovative and modern video games. They became popular outside Japan with the video game Space Invader launched in 1978, followed by Pac-Man in 1980. What helped Japan excel in the video game markets despite the lack of years of experience in this field was mainly thriving four other industries that coincided with the entry of video games, namely 1) electrical equipment, 2) computer, 3) television, and 4) mass production mainly found in the Japanese automobile industry, motorcycles, and textiles. The principles, inputs, and mechanisms of action of these four industrial fields intersected with the video game industry (197).

The Japanese company Nintendo introduced the Super Mario game in 1985, which is the best-selling game in the history of video games and has sold more than 48 million copies and generated US$ 2.5 billion (200, 201).

To analyze the video game industry in Japan and how technology and innovation subsequently passed on, we will sequence the historical events related to the beginning of import, imitation, and simulation in Japan after classifying video games into arcades (coin-operated entertainment video game machines) and household.

Four Japanese companies, namely Taito, Sega, Nintendo, and Namco, started importing the first arcade machines (Pong) from the USA. Taito, established in 1953, imported arcade machines in 1972. Before that, the company's activity was limited to cooperating with the US Walt Disney Company to import toys to Japan. Sega was established by merging two small companies, one of which was for US investors residing in Japan. The company has specialized in importing arcade video games from the US since 1972. Namco, established in 1955, specialized in manufacturing mechanical horses for children. In 1974, the US Atari Corporation decided to make Namco its distribution agent in Japan, given its extensive distribution and points of sale network (197).

Nintendo, founded in 1889 during the reign of Emperor Meiji, was the oldest of these companies. It manufactured card games and was famous for the Hanafuda card game, mechanical Carrom-like games called Pachinko, and board games. It began importing video games from 1973 to 1977. It changed its activity later. In 1973, Nintendo began importing the Laser Clay Shooting game, then importing a series of video games (197).

These four companies started their activities (or modified them later) by importing US video game products. Still, soon they began to produce a series of video games locally by copying the imported products, followed by the production of new video game products within two years. Still, they were not very popular in the market compared to the imported products. They did not have an active market outside Japan.

For example, Taito released its Elepong, a counterfeit Atari's Pong, in 1973, followed a few months later by Sega's Pong Tron 1, then Pong Tron 2, and then Hockey TV. Taito later released Soccer and Hockey Pro. These were all imitative products of the US companies (197, 202).

With the development of the innovative dose of these games, Taito exported the video game produced in 1974, namely Speed Race, and in 1975 it produced Western Gun. However, these games did not have a high export market until Taito produced the arcade video games Space Invaders. The success of this product decreased the 100-yen coins in the market due to the machine's dependence on it to operate, and more than 400,000 units were sold within four years. It achieved a legendary sales figure at the time of US$3.8 billion (a net profit of US$450 million) (197, 203, 204).

In 1977, Nintendo reproduced the successful video games in the US, such as Othello, Block Farer, and Space Fever. It then developed products, including Monkey Magic and Sheriff in1979. However, the release of Donkey Kong in 1981 was the beginning of success for Nintendo, as it sold more than 60,000 units in one year and gained US$280 million in sales in two years (1981-1983). (197, 205, 206).

Bandi Namco developed the Pac-Man in 1980 with sales of more than US$ one billion in one year. Even Atari took a license from Namco to resell and manufacture the product in the US. The technological cycle was reversed in technology transfer from Japan to the US in less than ten years (197, 207, 208).

As for home video games mentioned above, the US producers were among the first innovators in this field, namely Odyssey (1972) and Atari 2600 (1977). The Japanese companies focused on importing and then copying and imitating these products. The process took about ten years to reach innovation. However, Japanese companies could not excel until Nintendo released the Family Computer video game console in 1983 (197, 209).

Before that, many attempts were made, including the cooperation of the Japanese gaming company Epoch with Magnavox [producer of the Odyssey video game console in 1972 and later part of Philips] to produce a locally made version of the Epoch TV Tennis game in Japan in 1975, which sold three million copies. Then major Japanese television companies, such as Toshiba, Sharp, and Panasonic, began reproducing home video game products. Namco also released the 1000 TV Jack in

1977 and the Super Vision game in 1979; Toshiba released the Visicorn in 1978 (197, 210).

⌘

During two decades, from the early 1980s to the early 2000s, Japan produced and exported the most video games of all kinds. The US did not regain competition until Microsoft released its X-Box in 2002 (211).

The video game industry in Japan was accompanied by the development of the personal computer industry, as Japanese companies specialized in electronics, such as Sony, Hitachi, Canon, and Mitsubishi Electric, imported microprocessors from the USA, such as Sharp MZ-80 k and Intel 8080 in the 1970s and early 1980s and produced personal computers based on the US microprocessor, such as PC-8001 (1972), PC-8801 (1981), PC-980 (1982), Sharp X1 (1982), and Fujitsu FM-7 (1982). Then Japanese companies developed processors, innovated, and competed internationally (212-214).

⌘

References

1. J. Habu, *Ancient Jomon of Japan*. (Cambridge University Press, Cambridge, UK; New York, NY, USA, 2004).

2. W. Hong, Yayoi Wave, Kofun Wave, and Timing: The Formation of the Japanese People and Japanese Language. *Korean Studies* **29**, 1-29 (2005).

3. H. Kanaseki, M. Sahara, The Yayoi Period. *Asian Perspectives* **19**, 15-26 (1976).

4. K. Misaka, K. Wakabayashi, in *Coexistence and Cultural Transmission in East Asia*, W. A. C. Inter-Congress, N. Matsumoto, H. Bessho, M. Tomii, Eds. (Left Coast Press, 2011).

5. S. Turnbull, W. Reynolds, *Fighting Ships of the Far East (2): Japan and Korea AD 612–1639*. (Bloomsbury Publishing, 2012).

6. C. D. Benn, C. Benn, *China's Golden Age: Everyday Life in the Tang Dynasty*. (Oxford University Press, 2004).

7. M. E. Lewis, *China's Cosmopolitan Empire: The Tang Dynasty*. (Harvard University Press, 2012).

8. J. Russell, R. Cohn, *Taika Reform*. (Book on Demand, 2012).

9. C.-Y. Liao, University of Canterbury (2006).

10. B. L. Batten, Foreign Threat and Domestic Reform: The Emergence of the Ritsuryō State. *Monumenta Nipponica* **41**, 199-219 (1986).

11. The Japanese Watch & Clock Industry. (JWCI, 2020).

12. D. Fléchon, F. Cologni, F. d. l. h. horlogerie, *The Mastery of Time: A History of Timekeeping, from the Sundial to the Wristwatch: Discoveries, Inventions, and Advances in Master Watchmaking*. (Flammarion, 2011).

13. R. R. J. Rohr, *Sundials: History, Theory, and Practice*. (Dover Publications, 2012).

14. W. E. Marks, in *Water Encyclopedia*. pp. 704-707.

15. R. T. Balmer, The invention of the sand clock. *Endeavour* **3**, 118-122 (1979).

16. D. Deming, *Science and Technology in World History, Volume 2: Early Christianity, the Rise of Islam and the Middle Ages.* (McFarland, Incorporated, Publishers, 2014).

17. J. Hawkins, *A sartorial tale: Evening wear for men: The style and the times.* (Antiques & Art in Australia, 2013), pp. 18–25.

18. Y. Yokota. (Springer Netherlands, Dordrecht, 2009), pp. 175-188.

19. I. Shoji, Technology Transfer from Japanese to Indian Firms. *Economic and Political Weekly* **20**, 2031-2042 (1985).

20. T. Hashimoto, Kansei Calendar, Japanese clocks and the definition of twilight. *Astronomical Herald* **98**, 373 (2005).

21. Y. Masao, in *6th:; International studies conference, European Association for Japanese Studies; Japan at play the ludic and logic of power.* (Routledge,;, London:, Berlin, 2005), pp. 72-83.

22. D. M. Brown, The Impact of Firearms on Japanese Warfare, 1543-98. *The Far Eastern Quarterly* **7**, 236-253 (1948).

23. N. Perrin, *Giving Up the Gun: Japan's Reversion to the Sword, 1543-1879.* (D. R. Godine, 1979).

24. K. Chase, K. W. Chase, *Firearms: A Global History to 1700.* (Cambridge University Press, 2003).

25. O. G. Lidin, N. I. o. A. Studies, *Tanegashima: The Arrival of Europe in Japan.* (Nordic Institute of Asian Studies, 2002).

26. B. J. Buchanan, *Gunpowder, Explosives and the State: A Technological History.* (Ashgate, 2006).

27. M. Adas, *Technology and European Overseas Enterprise: Diffusion, Adaptation and Adoption.* (Routledge, 1996).

28. S. R. Turnbull, M. Boxall, *Samurai Warfare.* (Arms and Armour Press, 1996).

29. G. Feifer, *Breaking Open Japan: Commodore Perry, Lord Abe, and American Imperialism in 1853*. (HarperCollins, 2013).

30. F. L. Hawks, *Commodore Perry's Expedition to Japan*. (Createspace Independent Pub, 2013).

31. F. L. Hawks, *Commodore Perry and the Opening of Japan: Narrative of the Expedition of an American Squadron to the China Seas and Japan, 1852-1854 : the Official Report of the Expedition to Japan*. (Nonsuch, 2005).

32. J. L. Coleman, The American whale oil industry: A look back to the future of the American petroleum industry? *Nonrenewable Resources* **4**, 273-288 (1995).

33. U. S. N. Archives, R. Administration, *The Treaty of Kanagawa: Setting the Stage for Japanese-American Relations*. (National Archives and Records Administration, 2003).

34. R. Blumberg, *Commodore Perry in the Land of the Shogun*. (HarperCollins, 2003).

35. C. D. Totman, *Tokugawa Ieyasu, Shogun: A Biography*. (Heian, 1983).

36. A. Gordon, *A Modern History of Japan: From Tokugawa Times to the Present*. (Oxford University Press, 2014).

37. C. Adele, *A Guide to the Japanese Civil War: The Boshin War Of 1868*. (BiblioBazaar, 2012).

38. R. G. Flershem, Some Aspects of Japan Sea Shipping and Trade in the Tokugawa Period, 1603-1867. *Proceedings of the American Philosophical Society* **110**, 182-226 (1966).

39. T. C. Smith, The Introduction of Western Industry to Japan During the Last Years of the Tokugawa Period. *Harvard Journal of Asiatic Studies* **11**, 130-152 (1948).

40. P. Davies, in *Chapter 7. Japanese Shipping And Shipbuilding: An Introduction To The Motives Behind Its Early Expansion*. (Brill, 2010), pp. 114-120.

41. S. Crawcour, The Tokugawa Period and Japan's Preparation for Modern Economic Growth. *Journal of Japanese Studies* 1, 113-125 (1974).

42. L. Books, G. B. LLC, *1869 in Japan: Boshin War, Battle of Hakodate, Republic of Ezo, Naval Battle of Miyako Bay*. (General Books LLC, 2010).

43. W. Beasley, *The Meiji Restoration*. (Stanford University Press, 1972).

44. H. CORTAZZI, *Britain & Japan: Biographical Portraits, Vol. X*. H. Cortazzi, Ed., Britain & Japan: Biographical Portraits, Vol. X (Renaissance Books, 2016), vol. 10.

45. B. Micou, Torpedo Boats in Naval Warfare. *The North American Review* **165**, 409-417 (1897).

46. K. Yamamura, Success Illgotten? The Role of Meiji Militarism in Japan's Technological Progress. *The Journal of Economic History* **37**, 113-135 (1977).

47. J. C. Perry, Great Britain and the Emergence of Japan as a Naval Power. *Monumenta Nipponica* **21**, 305-321 (1966).

48. J. C. Schencking, The Imperial Japanese Navy and the Constructed Consciousness of a South Seas Destiny, 1872-1921. *Modern Asian Studies* **33**, 769-796 (1999).

49. S. Broadbridge, Shipbuilding and the State in Japan since the 1850s. *Modern Asian Studies* **11**, 601-613 (1977).

50. P. Olender, *Sino-Japanese Naval War 1894-1895*. (Mushroom Model Publications, 2014).

51. S. C. M. Paine, *The Sino-Japanese War of 1894–1895: Perceptions, Power, and Primacy*. (Cambridge University Press, 2002).

52. H.-C. Wang, Discovering Steam Power in China, 1840s-1860s. *Technology and Culture* **51**, 31-54 (2010).

53. H.-c. Wang, in *Encyclopaedia of the History of Science, Technology, and Medicine in Non-Western Cultures,* H. Selin, Ed. (Springer Netherlands, Dordrecht, 2008), pp. 1-6.

54. F. Wakeman, Rebellion and Revolution: The Study of Popular Movements in Chinese History. *The Journal of Asian Studies* **36**, 201-237 (1977).

55. D. J. Silbey, *The Boxer Rebellion and the Great Game in China: A History.* (Farrar, Straus and Giroux, 2012).

56. G. Jukes, *The Russo-Japanese War 1904–1905.* (Bloomsbury Publishing, 2014).

57. I. H. Nish, P. Longman, *The Origins of the Russo-Japanese War.* (Longman, 1985).

58. J. Steinberg *et al.*, *The Russo-Japanese War in Global Perspective: World War Zero, Volume II.* (Brill, 2006).

59. J. Corbett, *Maritime Operations in the Russo-Japanese War, 1904?1905: Volume 1.* (Naval Institute Press, 2015).

60. S. Tan, Race in the Shipbuilding Industry: Cases of South Korea, Japan and China. *International Journal of East Asian Studies* **6**, 65-81 (2017).

61. S. Tenold, in *Shipping and Globalization in the Post-War Era. Palgrave Studies in Maritime Economics,* Petersson N., Tenold S., W. N., Eds. (Palgrave Macmillan, 2019).

62. J. Hunter, H. Macnaughtan, *National Histories of Textile Workers: Japan 1650-2000.* (Ashgate Publishing Ltd, 2010).

63. Kimonoboy. (2021), vol. 2021.

64. A. S. Pearse, The Cotton Industry of Japan, China and India and Its Effect on Lancashire. *International Affairs (Royal Institute of International Affairs 1931-1939)* **11**, 633-657 (1932).

65. K. Tamagawa, paper presented at the The 1st International Conference on Business & Technology Transfer (ICBTT 2002), Kyoto, Japan, 2002.

66. M. Miyamoto. (1989).

67. G. Saxonhouse, A Tale of Japanese Technological Diffusion in the Meiji Period. *The Journal of Economic History* **34**, 149-165 (1974).

68. S. Ichimura, in *Political Economy of Japanese and Asian Development.* (Springer Japan, 2013).

69. *The History of U.S. Cotton in Japan* (JA0503, 2010).

70. J. Hunter, paper presented at the Ninth conference of the Global Economic History Network, Kaohsiung, Taiwan, 2006.

71. T. Nakaoka, K. Aikawa, H. Miyajima, T. Yoshii, T. Nishizawa, The Textile History of Nishijin (Kyoto): East Meets West. *Textile History* **19**, 117-141 (1988).

72. D. E. Westney, *Imitation and Innovation.* (Harvard University Press, 2013).

73. A. D. Brown, Meiji Japan: A unique Technological Experience? *Student Economic Review* **19**, 71-83 (2005).

74. N. Ike, Western Influences on the Meiji Restoration. *Pacific Historical Review* **17**, 1-10 (1948).

75. *Encouragment of Invention in Japan* (2009).

76. T. Nicholas, H. Shimizu, Intermediary Functions and the Market for Innovation in Meiji and Taishō Japan. *The Business History Review* **87**, 121-149 (2013).

77. Japanese Patents and Inventions. *Nature* **135**, 218-218 (1935).

78. T. Nicholas, The origins of Japanese technological modernization. *Explorations in Economic History* **48**, 272-291 (2011).

79. T. Nicholas, HYBRID INNOVATION IN MEIJI, JAPAN. *International Economic Review* **54**, 575-600 (2013).

80. Z. Hotta, in *US916869A*. (United States, 1909).

81. Benrishi. (2013), vol. 2021.

82. Y. Huang, J. Zhang, S. J. T. U. Press, *Materials Corrosion and Protection.* (De Gruyter, 2018).

83. W. V. S. Baeckmann, W.; Prinz, W. , in *Handbook of Cathodic Corrosion Protection*, W. V. Baeckmann, Schwenk, W., , Ed. (Elsevier, Houston, TX, USA, 1997).

84. M. Shintani, TECHNOLOGICAL PROGRESS IN THE TEA MANUFACTURING INDUSTRY IN JAPAN. *Hitotsubashi Journal of Economics* **32**, 21-38 (1991).

85. K. S. Kardam, "Utility model –A tool for economic and technological development: A case study of Japan " (Indian Patent Office, New Delhi, INDIA 2007).

86. T. Nicholas, The Organization of Enterprise in Japan. *The Journal of Economic History* **75**, 333-363 (2015).

87. W. Röhl, *History of law in Japan since 1868* (Brill, Leiden 2005).

88. M. D. Burton, T. Nicholas, "Prizes, Patents and the Search for Longitude," (Harvard Business School, 2016).

89. R. Burrell, C. Kelly, PARLIAMENTARY REWARDS AND THE EVOLUTION OF THE PATENT SYSTEM. *The Cambridge Law Journal* **74**, 423-449 (2015).

90. M. Daniel, Daguerre (1787–1851) and the Invention of Photography. *Heilbrunn Timeline of Art Histor.* 2004.

91. G. A. Wickliff, The Daguerreotype and the Rhetoric of Photographic Technology. *Journal of Business and Technical Communication* **12**, 413-436 (1998).

92. M. S. Barger, S. M. Barger, W. E. White, W. B. White, Smithsonian, *DAGUERREOTYPE.* (Smithsonian, 1991).

93. A. Goto *et al.*, *Innovation in Japan.* (Clarendon Press, 1997).

94. D. Comin, B. Hobijn, An Exploration of Technology Diffusion. *The American Economic Review* **100**, 2031-2059 (2010).

95. K. Akamatsu, A Theory of Unbalanced Growth in the World Economy. *Weltwirtschaftliches Archiv* **86**, 196-217 (1961).

96. K. Akamatsu, A HISTORICAL PATTERN OF ECONOMIC GROWTH IN DEVELOPING COUNTRIES. *The Developing Economies* **1**, 3-25 (1962).

97. M. Lipton, Balanced and Unbalanced Growth in Underdeveloped Countries. *The Economic Journal* **72**, 641-657 (1962).

98. M. E. Porter, M. Sakakibara, Competition in Japan. *Journal of Economic Perspectives* **18**, 27-50 (2004).

99. S. Parissien, *The Life of the Automobile: The Complete History of the Motor Car.* (St. Martin's Publishing Group, 2014).

100. S. J. C. Nixon, *The Invention of the Automobile - (Karl Benz and Gottlieb Daimler).* (Edizioni Savine, 2016).

101. E. Eckermann, *World History of the Automobile.* (Society of Automotive Engineers, 2001).

102. M. Walker, *Motorcycle: Evolution; Design; Passion.* (Octopus Publishing Group, 2014).

103. J. W. Alexander, *Japan's Motorcycle Wars: An Industry History.* (UBC Press, 2008).

104. E. Yamamura, T. Sonobe, K. Otsuka, Time path in innovation, imitation, and growth: the case of the motorcycle industry in postwar Japan. *Journal of Evolutionary Economics* **15**, 169-186 (2005).

105. Honda Motorcycles. (Honda Motorcycles, 2021), vol. 2021.

106. Suzuki. (Suzuki, 2021), vol. 2021.

107. J. A. Dodge, Joseph Priestley. *Journal of the Minnesota Academy of Science* **2**, 228-238 (1881).

108. D. Steen, Carbonated beverages. *Chemistry and technology of soft drinks and fruit juices*, 150-180 (2005).

109. W. A. Smeaton, Torbern Olof Bergman: from natural history to quantitative chemistry. *Endeavour* **8**, 71-74 (1984).

110. M. E. Weeks. (ACS Publications, 1957).

111. A. Lennartson, in *Carl Wilhelm Scheele and Torbern Bergman*. (Springer, 2020), pp. 235-244.

112. E. Bly, Just what the doctor ordered: a medical history of soft drinks. (2007).

113. J. Chandler, Edgeworth and the Lunar Enlightenment. *Eighteenth-Century Studies*, 87-104 (2011).

114. A. C. Funderburg, *Sundae Best: A History of Soda Fountains*. (Bowling Green State University Popular Press, 2002).

115. G. B. Roberts, Dr. Physick and His House. *The Pennsylvania Magazine of History and Biography* **92**, 67-86 (1968).

116. T. Donovan, *Fizz: How soda shook up the world*. (Chicago Review Press, 2013).

117. F. Ebner, James Vernor. *Journal of the American Pharmaceutical Association* **8**, 529-529 (1919).

118. E. Fasoli, A. D'Amato, A. Citterio, P. G. Righetti, Ginger Rogers? No, ginger ale and its invisible proteome. *Journal of proteomics* **75**, 1960-1965 (2012).

119. D. S. M. Lewis, *Moxie*. (Arcadia Publishing, 2019).

120. S. Fraser, Labor's True Women: Carpet Weavers, Industrialization, and Labor Reform in the Gilded Age. By Susan Levine.(Philadelphia, Pa.: Temple University Press, 1984. xii+ 191 pp. Illustrations, tables, notes, and index. $24.95.). *Business History Review* **61**, 144-146 (1987).

121. J. Baumer, *Moxie: Maine in a Bottle*. (Down East Books, 2012).

122. V. Bodden, *The Story of Coca-Cola*. (Creative Education, 2009).

123. H. McQueen, The essence of modernity. *Island*, 15-19 (1992).

124. J. H. Young, Three Atlanta Pharmacists. *Pharmacy in history* **31**, 16-22 (1989).

125. F. Gulzar, Brand Failures-When do Good Brands do Bad."(A Case Study of Classic Example of Brand Failure: Pepsi Cafe Chino). *KHOJ: Journal of Indian Management Research and Practices*, 94-101 (2015).

126. S. G. Llanas, *Caleb Davis Bradham: Pepsi-Cola Inventor: Pepsi-Cola Inventor.* (Abdo Publishing, 2014).

127. M. W. Martin, *Twelve Full Ounces.* (Holt, Rinehart and Winston, 1962).

128. D. Bridgforth, *Mountain Dew: The History.* (Booksurge, 2007).

129. S. Asaba, M. B. Lieberman, *Why Do Firms Behave Similarly?: A Study on New Product Introduction in the Japanese Soft-drink Industry.* (Center on Japanese Economy and Business, Columbia Business School, 1999).

130. R. A. D'Aveni, Waking up to the new era of hypercompetition. *The Washington Quarterly* **21**, 183-195 (1998).

131. D. B. Yoffie, and Yusi Wang., Cola Wars Continue: Coke and Pepsi in the Twenty-First Century. *Harvard Business School Case* **702-442**, (2002).

132. ReportLinker, "Global Soft Drinks Industry," (ReportLinker, 2021).

133. B. T. Mitsui, The System of Communications at the Time of the Meiji Restoration. *Monumenta Nipponica* **4**, 88-101 (1941).

134. Y. SHIRAHATA, The modernization of transport and communications in Japan. *International Research Center for Japanese Studies*, 75-86 (1998).

135. D. Sablosky, (2016).

136. V. Cârstea, Delocalizing the Japanese automotive industry and the Romanian market. *Romanian Economic and Business Review* **10**, 278 (2015).

137. H. T. Hlynsson, (2020).

138. W. James, *Driving from Japan: Japanese Cars in America.* (McFarland, Incorporated, Publishers, 2015).

139. A. Besher, J. Wilcock, *The Pacific Rim Almanac.* (HarperPerennial, 1991).

140. B. Laban, N. Baldwin, G. N. Georgano, *The World Guide to Automobile Manufacturers.* (Facts on File Publications, 1987).

141. D. D. Dauphinais, P. M. Gareffa, *Car Crazy: The Official Motor City High-octane, Turbocharged, Chrome-plated, Back Road Book of Car Culture.* (Visible Ink Press, 1996).

142. H. Kumon, in *Japanese Foreign Direct Investment and the East Asian Industrial System.* (Springer, 2002), pp. 337-351.

143. J. Begley, T. Donnelly, The DaimlerChrysler Mitsubishi merger: a study in failure. *International journal of automotive technology and management* **11**, 36-48 (2011).

144. C. S. Chang, *THE JAPANESE MOTOR VEHICLE INDUSTRY: A STUDY OF THE HISTORY OF THE JAPANESE MOTOR VEHICLE INDUSTRY AND THE IMPACT OF THE JAPANESE MOTOR VEHICLES ON THE UNITED STATES MARKET.* (American University, 1974).

145. J. Berengueres, *The Toyota production system re-contextualized.* (Lulu. com, 2007).

146. M. A. Cusumano, Manufacturing innovation: lessons from the Japanese auto industry. *MIT Sloan Management Review* **30**, 29 (1988).

147. M. A. Cusumano, The Japanese automobile industry: Technology and management at Nissan and Toyota. (1985).

148. M. Udagawa, The Prewar Japanese Automobile Industry and American Manufacturers. *Japanese Yearbook on Business History* **2**, 81-99 (1986).

149. K. Togo, Infant Industry Policy: A Case of Japanese Automobile Industry Before 1945. *Available at SSRN 963953*, (2007).

150. R. Cox, *The Culture of Copying in Japan: Critical and Historical Perspectives.* (Taylor & Francis, 2007).

151. M. Takehara, N. Hasegawa, in *Sustainable Management of Japanese Entrepreneurs in Pre-War Period from the Perspective of SDGs and ESG.* (Springer, 2020), pp. 127-143.

152. H. Ueno, H. Muto, in *Industry and Business in Japan.* (Routledge, 2017), pp. 139-190.

153. S. C. Townsend, The 'miracle'of car ownership in Japan's 'Era of High Growth', 1955–73. *Business History* **55**, 498-523 (2013).

154. C. J. Singleton, Auto industry jobs in the 1980's: a decade of transition. *Monthly Labor Review* **115**, 18-27 (1992).

155. C. Lin, The japanese automotive industry: recent developments and future competitive outlook. (1994).

156. A. NICOLE, Imported From Detroit. *Jacobin*, (2013).

157. S.-. Yoshimi, Made in Japan': the cultural politics ofhome electrification'in postwar Japan. *Media, Culture & Society* **21**, 149-171 (1999).

158. R. N. Langlois, W. E. Steinmueller, Strategy and circumstance: The response of American firms to Japanese competition in semiconductors, 1980–1995. *Strategic Management Journal* **21**, 1163-1173 (2000).

159. W. F. Brinkman, D. E. Haggan, W. W. Troutman, A history of the invention of the transistor and where it will lead us. *IEEE Journal of Solid-State Circuits* **32**, 1858-1865 (1997).

160. M. Riordan, L. Hoddeson, C. Herring, The invention of the transistor. *More Things in Heaven and Earth*, 563-578 (1999).

161. J. Gertner, *The idea factory: Bell Labs and the great age of American innovation.* . (Penguin, 2012).

162. H. Choi, O. Takushi, Failure to launch: Tarui Yasuo, the quadrupole transistor, and the meanings of the IC in postwar Japan. *IEEE Annals of the History of Computing* **34**, 48-59 (2011).

163. D. Brock, L. David, The early history of microcircuitry: An overview. *IEEE Annals of the History of Computing* **34**, 7-19 (2011).

164. M. Watanabe, Semiconductor industry in Japan—past and present. *IEEE Transactions on Electron Devices* **31**, 1562-1570 (1984).

165. A. Morita, E. M. Reingold, M. Shimomura, M. Smith, *Made in Japan: Akio Morita and Sony.*, (Dutton, New York, 1986).

166. K. Sakui, Professor Fujio Masuoka？ s Passion and Patience Toward Flash Memory. *IEEE Solid-State Circuits Magazine* **5**, 30-33 (2013).

167. USJC. (USJC, 2021), vol. 2021.

168. K. Sakakibara, From imitation to innovation: the very large scale integrated (VLSI) semiconductor project in Japan. *Sloan School of Management* (1983).

169. L. Lynn, Japanese robotics: Challenge and—limited—exemplar. *The Annals of the American Academy of Political and Social Science* **470**, 16-27 (1983).

170. Orange Labs Tokyo, «Japan Robot Market Overview,» (2016).

171. M. Shelley, *Frankenstein.* (Dover Publications, 2013).

172. D. Fishelson, J. C. Landis, D. P. Service, *The Golem.* (Dramatists Play Service, 2001).

173. H. M. Collins, T. Pinch, *The Golem: What You Should Know About Science.* (Cambridge University Press, 2012).

174. E. S. Ellis, *The Steam Man of the Prairies.* (Dover Publications, 2016).

175. L. F. Baum, *Tik-Tok of Oz.* (Dover Publications, 2012).

176. L. F. Baum, *Ozma of Oz*. (Platanus Publishing, 2020).

177. K. Capek, C. Novack-Jones, I. Klima, *R.U.R. (Rossum's Universal Robots)*. (Penguin Publishing Group, 2004).

178. J. Sobota, J. Sobota, *I, Robot: Three Laws of Robotics*. (Jan & Jarmila Sobota, 2007).

179. W. L. Stone, in *Robotics and Automation Handbook*. (CRC Press, 2018), pp. 8-19.

180. G. C. Devol, U. P. office, Ed. (United States, 1961).

181. J. F. Engelberger, I. Asimov, D. Lock, K. Willis, *Robotics in Practice: Management and Applications of Industrial Robots*. (AMACOM, 1980).

182. B. Singh, N. Sellappan, P. Kumaradhas, Evolution of industrial robots and their applications. *International Journal of emerging technology and advanced engineering* **3**, 763-768 (2013).

183. C. Laschi *et al.*, Grasping and manipulation in humanoid robotics. *Italia*, (2000).

184. S. Kajita, H. Hirukawa, K. Harada, K. Yokoi, *Introduction to humanoid robotics*. (Springer Berlin Heidelberg, 2014).

185. O. Wallach, in *Visualcapitalist*. (Visualcapitalist, 2020).

186. W. Higinbotham, "Tennis for two," *Analog computer/oscilloscope game* (Brookhaven National Laboratory, 1958).

187. M. O'Hagan, Putting pleasure first: localizing Japanese video games. *TTR: traduction, terminologie, rédaction* **22**, 147-165 (2009).

188. P. Morrison. (JSTOR, 1980).

189. N. Diak. (Taylor & Francis, 2019).

190. A. Abdela, J. E. Weis, V. Friederici, B. Verreet, P. Westbergh, FUTURE DISPLAY TECHNOLOGIES.

191. J. Herz, in *What Higher Education Can Learn from Multiplayer O nline Worlds. In R. LM D evlin, Internet and the U niversity: 2001 Forum.* (2002), pp. 169-191.

192. M. Demirbilek, in *Encyclopedia of Distance Learning, Second Edition.* (IGI Global, 2009), pp. 2209-2223.

193. E. Huhtamo, Slots of fun, slots of trouble. *The handbook of computer game studies. Cambridge, MA: MIT Press [Google Scholar],* (2005).

194. M. Z. Newman, *Atari age: The emergence of video games in America.* (MIT Press, 2017).

195. J. Bedi, Ralph Baer: An interactive life. *Human Behavior and Emerging Technologies* **1**, 18-25 (2019).

196. A. Wilcox, Regulating Violence in Video Games: Virtually Everything. *J. Nat'l Ass'n Admin. L. Judiciary* **31**, 253 (2011).

197. M. Picard, The foundation of geemu: A brief history of early Japanese video games. *Game Studies* **13**, (2013).

198. M. Koizumi, in *Transnational Contexts of Development History, Sociality, and Society of Play.* (Springer, 2016), pp. 13-64.

199. M. Consalvo, Convergence and globalization in the Japanese videogame industry. *Cinema Journal* **48**, 135-141 (2009).

200. J. Togelius, S. Karakovskiy, J. Koutník, J. Schmidhuber, in *2009 ieee symposium on computational intelligence and games.* (IEEE, 2009), pp. 156-161.

201. Vgsales. (Vgsales, 2021), vol. 2021.

202. M. J. Wolf, *The video game explosion: a history from PONG to Playstation and beyond.* (ABC-CLIO, 2008).

203. I. Towers, L. Duxbury, C. Higgins, J. Thomas, Time thieves and space invaders: Technology, work and the organization. *Journal of Organizational Change Management,* (2006).

204. D. R. Ross, D. H. Finestone, G. K. Lavin, Space invaders obsession. *JAMA* **248**, 1177-1177 (1982).

205. M. J. Wolf, Genre and the video game. *The medium of the video game* **1**, 113-134 (2001).

206. K. Horowitz, *Beyond Donkey Kong: A History of Nintendo Arcade Games*. (McFarland, 2020).

207. R. F. Bowman, A" Pac-Man" theory of motivation: Tactical implications for classroom instruction. *Educational technology* **22**, 14-16 (1982).

208. M. Gallagher, A. Ryan, in *The 2003 Congress on Evolutionary Computation, 2003. CEC'03.* (IEEE, 2003), vol. 4, pp. 2462-2469.

209. D. Sheff, *Game over: How Nintendo conquered the world.* (Vintage, 2011).

210. JETRO, "Japanese Video Game Industry," (Japan External Trade Organization, 2007).

211. R. Marshall, The history of the Xbox. *Accessed online https://www. digitaltrends. com/gaming/the-history-of-the-xbox/. Viewed* **1**, 2017 (2013).

212. M. Fransman, *Japan's Computer and Communications Industry: the evolution of industrial giants and global competitiveness.* (Clarendon Press, 1995).

213. T. E. Stuart, J. M. Podolny, Local search and the evolution of technological capabilities. *Strategic management journal* **17**, 21-38 (1996).

214. S. Mandiberg, *Responsible localization: Game translation between Japan and the United States.* (University Of California, San Diego, 2015).

PART TWO

IMITATION AND
INNOVATION IN KOREA

CHAPTER EIGHT

CHAPTER EIGHT

FROM IMITATION TO INNOVATION IN KOREA

In Korea, per capita income rose from US$ 160 in 1960 to more than US$ 32,000 in 2020, nearly 200 times (1,2). Korea was one of the ten largest global economies in 2006, as per the GDP of about US$ 1 trillion though it does not have rich natural resources (3). Its total exports increased from nearly US$ two billion in 1960 to US$ 557 billion in 1996, i.e., 278 times (4).

How could Korea, which was under Japanese occupation from 1910 to 1945 (5, 6), then the US occupation and the civil war (1950-1953) (7, 8), which fragmented Korea, destroyed its resources and infrastructure, and killed nearly three million Koreans, bridge the technological gap with the developed countries that occupied it and became an innovative country that exports its products, especially electronics, cars, and ships, to the ends of the world?

The answer to this question is the focus of this chapter, which analyzes imitation and innovation during this period to identify its causes and stages. It is worth noting that the period of technological absorption in Korea was short, and the transition from imitation to innovation was fast, thus rendering the Korean model unique, especially when analyzing reliance on the national research system to feed innovative local content.

         ⋙

Although researchers have established various period divisions for Korea's transition from a developing to a developed country, these divisions differ per each author's point of view and the underlying factors of such divisions. Some divided it in terms of the volume of exports and imports; others divided it based on the regulations and legislation; a third group divided it as per the political chain of command; let alone the volume of foreign direct investment (FDI) and its role in the Korean economy (9-13).

However, in this book, we have a different division that intersects and overlaps with the aforementioned previous ones but with the underlying rationale of a nation not possessing certain technologies, passing by its ability to copy, imitate, and simulate the innovations of other countries and then develop modern innovations on its own.

Accordingly, we found it more appropriate to set 1910-1960 as the period in which Korea did not have modern innovations, then 1961-1989 for the start of copying and imitation, which in turn can be re-divided into three sub-periods: 1960-1969, 1970-1979, and 1980-1989. Then Korea started the era of innovation in the early 1990s. The bottom line is that Korea has gone through three major periods concerning innovation.

❧

In the first period before 1961, Korea was under Japanese occupation, the US influence, and civil war for fifty years even after independence in 1953 during the era of the First Republic headed by Syngman Rhee (1948-1960) (14). From 1960 to 1962, four presidents were overthrown, ending with the military coup on March 24, 1962, led by Park Chung-hee, who started what was known as the Third Korean Republic and ruled the country for nearly 17 years until his assassination in 1979 (15). The precursors of transition to imitation coincided with the rule of Park Chang-hee. The two years preceding his rule witnessed severe political fluctuations under four presidents. Knowing the style and methodology of the country's political leadership may lead us to identify the source of legislation, strategic directions, and visions for the nation. The transition to innovation (as we have observed in other countries) stems from the legislative authority of the nation, accompanied by interdependence and overlap between the nation's institutions, the response of its society, then the country's relations with innovative nations and how to adopt the simulation and copying methodology of foreign innovations in the early periods. Accordingly, political authority is the source of legislation and the steering wheel of transformation into an innovative nation.

The analysis of the period before 1961 shows that Korea did not possess modern technologies or advanced industries and could not even simulate or copy any innovations reaching its ports except for what was established by the Japanese government during its colonization of Korea after the 1910 Japan–Korea Annexation Treaty, which placed Korea under Japan's protectorate. Companies and factories in Korea under occupation were established per the Japanese style. In previous chapters, we explained how Japan became an innovative country in the modern era. Japan transferred some industries to Korea, especially the railway industry, the banking industry, and the holding company system. Japan began building a railway linking the Chun Zechon Port to the Korean capital between 1900-1910 to pave the way for a quick outlet for the military forces after landing in the port as it became clear after the occupation. It was to repel any attack or expansion of China in the region (16-18).

The precursors of occupation were moving foreign investments to Korea, intensifying these interests, and then demanding to protect them, ensure their safety, and enhance their success. Just as foreign investments sometimes contribute to the renaissance and growth of countries, they may be a reason to invade them and a pretext for occupying and placing them under protection.

❧

With the beginning of the second period (copying and imitation) in 1961, Korea implemented the "Import Substitution Policy" by enacting laws and setting incentive programs that encourage industrialization of everything imported and setting up disincentive programs to reduce imports. A series of restrictions on imported products was initiated in favor for a domestic manufacture of a particular product. In order for Korea to reach this stage in the local industry, it resorted to two measures, namely foreign technological licenses and foreign technological consultations. Through technological licenses, Korean factories transferred technology to copy, simulate, and imitate the imported foreign products, and through foreign technological consultations, it could train, teach, maintain, and repair devices and equipment imported for local manufacturing, which was basically copies and imitations of what was imported, but with licenses from the technology owners. At this stage, Korea needed to build workforce capacity in technological and engineering fields. The process of importing technology required a minimum level of technological capabilities of the local workforce to benefit from the imported technology, and without it, countries cannot even begin copying and imitation, a process that requires skills and infrastructure that are not easy to prepare and qualify. Training and qualification of local workforce was key to deal with and operate imported technologies. Re-manufacturing imported products within the country requires support from the country's government along with protection by discouraging and restricting the import of products so that the national entities can compete in the beginning (19-21).

The Korean government subsidized the domestic private sector. Foreign companies and branches of international companies move around the world looking for better markets or lower wages, and they are never a reliable source to generate sustainable growth for specific countries. The State's investment was directed to developing capabilities in the national private sector.

In 1962-1983, the Korean government imposed strict restrictions on foreign direct capital entry into the country to reduce and regulate foreign influence and ensure the development of industry and national investment with national capital, especially in the early stages. Some restrictions were later relaxed gradually in 1984-1994, and the activities/projects in which foreign direct capital was prohibited decreased to only 147 projects compared to the previous period. Then it began to stimulate foreign direct capital in some fields, starting in 1995. FDIs increased from US$ 1.36 billion in 1995 to US$ 11 billion in 1999, nearly nine times in just four years after the policy concerning foreign investment changed (19).

The benefits of these restrictions on foreign investments are evident today. They protected the nascent national industries and investments. They were followed by gradual easing in each industrial sector and activity according to the ability of the local industry to grow in simulation, imitation, and copying. The efforts also focused on turn-key factories not specialized in producing a specific industry but used in several industries to keep pace with supply and demand and the transformation in economic activities, especially during the first period of economic growth.

❧

As an intermediate stage of progress in industrial growth and economic prosperity, Korea shifted from the "Import Substitution" Policy to the "Export Promotion" Policy in the 1970s. There was a shift in production volume first and then production quality. The goal in the previous stage (Import Substitution) was to create a local industry capable of substituting the imported goods. The Korean government's direct support accompanied this endeavor on the one hand, along with the restriction on imported products in favor of national products, which were imitations and of poor quality, compared to the foreign imported product. However, this created a local economy capable of covering basic industrial needs and correcting the trade balance between Korea and foreign exporting countries, which were depleting hard currency to bring foreign products into the country (13, 22).

The goal of the "Export Promotion" phase (an advanced stage of copying, imitation, and simulation) was to bring in foreign currency from abroad and correct the trade balance in favor of Korea. However, this phase was not easy, as the Korean products in 1970-1980 were counterfeit of poor quality compared to products in developed countries. Korea's wager relied on two factors, namely the low-waged Korean labor, which makes the cost of the product lower by a profitable difference, making it reach the markets of other foreign countries. Furthermore, government generously supported the local industries through programs, regulations, and legislation that enabled the local exporting producer (investor) to receive benefits, financing, grants, subsidies, and tax exemptions.

In addition to the low-quality and low-cost Korean exports in 1960-1980 due to low-waged labor as one of the production inputs, the Korean exports were known in general for low-tech commodities, such as the shoes and textiles, which depended largely on low-cost labor. Thus, consumers had the impression that the Korean products were mere imitation, simple, and uncomplicated industries. The Korean government began to re-route its focus to exports with higher technological content (or higher added value). In 1960, exports constituted only 2% of the GDP. This percentage rose to 10% in 1970, and then 30% in 1980. However, 30% of these exports were shoes and textiles, which were simple and uncomplicated industries. However, the Korean government's contribution to programs to transform export quality increased the added value in export quality. Accordingly, the percentage of textiles and shoes decreased from 30% in 1960 to 10% in 1990 and then to less than 3% in 2000. As we mentioned at the beginning of this chapter about the revenues from total Korean exports, they rose from US$ two billion in 1960 to US$ 557 billion in 1996 (13), an indication of the change in the quality of Korean industries during the forty years between 1960 and 2000.

However, in the early 1980s, Korea received a technological shock because its industries and exports were based on imitation, simulation, and copying. In other words, if the licensing companies stops their licenses, Korean companies would not modernize their production lines and introduce developed industries. The technology licensing countries, such as Japan, the USA, and Europe sensed that the Korean industry could be competitive due to low prices and similar quality of the counterfeit product (in advanced stages of imitation and copying) and thus tried not to establish further assembly factories in Korea so as not to compete with them in the future. Examples included the refusal of the Japanese Mitsubishi Company to renew the technological auto engine license to the Korean Hyundai Company and the refusal of the US companies to sign technological licensing contracts with the Korean LG Company to manufacture color TV [we will discuss this in some detail in the next chapter].

Korea's spending on foreign technological transfer licenses rose from US$ 0.8 million in 1960-1966 to US$ one billion two hundred million in 1982-1988, i.e., more than 500 times. The spending on foreign technological consultancy increased from US$ 17 million in 1967-1971 to US$ 333 million in 1982-1986 (23).

Thus, research and innovation started to gain momentum. The desire for improving scientific and research capabilities resulted naturally from a natural sequence of events in the Korean industry whose hopes were pinned on the local research (the national research system) for its growth and overcoming its challenges towards innovation.

The government policy changed from encouraging exports to support the indigenous R&D policy at the end of the 1970s and early 1980s (1, 24).

Given the time lag between the inputs and outputs in scientific research projects and activities, it is not easy to time trace the development of the national research system based on outputs that may take a decade or more to impact the economy (25, 26), we will chronologically track some of the most prominent inputs to the activities contributed by the Korean government during the transition to a policy of supporting research and innovation. However, we assert that the outputs of these activities did not fructify until at least one to two decades later. The government focused on local knowledge stock whose development would enable the industry to progress and develop without relying entirely on imported foreign technologies, which were no longer available to Korean companies as they were before in the 1960s and 1970s for fear of competition. The local knowledge stock increased the capacity of local technological absorption and paved the way for developing imported technologies and producing innovations.

The government policies to support research and innovation in a bid to develop the local industry capacity comprised tax exemptions/relief on the income of research and innovation activities in 1975, income tax exemption up to 50% for researchers and engineers in 1980 (and later tax exemption on the income of profits from patents proceeds), tax exemptions for R&D based SMEs in 1985, the compensation system for investment losses for venture capital companies in 1980, the tax exemption on real estate lands containing R&D institutes in 1981, and the tax credit system to spend on developing the workforce in 1986 (27).

Other government policies included providing incentives and benefits to companies and the private sector as the main player in the innovation process, and keeping the government as a regulator and supporter. The government established Daedeok Innopolis as the first science park in Korea in 1979 in addition to direct government spending support for research. For comparison, government R&D support in 1963 was 1.2 billion Korean won, compared to 211 billion won in 1980, nearly 200 times in less than two decades. Then it rose from 211 billion Korean won in 1980 to 3.4 trillion won in 1990, i.e., 3,000 times in just two and a half decades [and later increased to 43 trillion won in 2010]. These figures prove the government's interest in research and innovation and its expected role in the country's future, especially after the shock of non-renewal of technological licenses. The percentage of spending on R&D of the GDP rose from 0.44% in 1976 to 1.59% in 1985 [which exceeded 3.7% later in 2010] (19, 24).

Given these government measures, full-time researchers increased from approximately 11,000 in 1976 to 100,000 in 1995, i.e., nine times during two decades, thus raising the number of researchers per 1,000 population from 0.33 to 2.24 in the same period [and then 15.4 in 2019] (27, 28).

The innovative output of the Korean research system increased from 50 patents filed with the USPTO for Koreans in 1980 to more than 5,000 patents filed in 2000, i.e., 100 times within twenty years (1).

The technological content (the added value of goods manufactured in Korea) increased over time. As a result, decades later, 23.4% of Korea's exports were classified as advanced technologies in 2000. About 68.8% of Korean exports were electronic exports (27).

Such technological progress was reflected in the average annual GDP growth, from 7.5% in the 1960s to 8.6% in the

1970s and 9.3% in the 1980s, i.e., twice Japan's economic growth of 4.5% and twice the US's of 3.2% (29).

Governmental measures of support and tax exemptions contributed to a gradual increase in the private sector's contribution in financing research, as the ultimate beneficiary of research and innovation outputs to improve production lines and industry in general. The private sector's contribution amounted to 81% of total spending (30).

Only 10% of R&D was carried out in universities, while the private sector and research institutions took over the rest. About two-thirds of researchers worked in the private sector. In 2000, Samsung spent more than US$ 500 billion on research and innovation, prompting a researcher to publish a chapter in a book entitled "The Samsung Republic." Its employees exceeded 287,000, and its annual profits amounted to US$ 121 billion. Recently, the percentage of Samsung spending on R&D exceeded government spending as a whole, as it reached 22%, compared to 20% of the government (23). Thus, that author who described the company as a republic did not exaggerate.

CHAPTER NINE

COPYING AND IMITATING THE INNOVATIONS OF ELECTRONICS, SHIPS, AND CARS IN KOREA

This chapter highlights three specific industrial imitation and innovation models in Korea: cars, shipbuilding, and electronics/semiconductor.

FROM IMITATION TO INNOVATION IN KOREAN VEHICLES

The Korean government played a significant and prominent role in the auto industry. It was not only the regulator of this sector but also the architect of policies and executive plans, the supporter of the private sector in this field, and its protector against foreign companies until it, at some stages of the industry's progress, prevented the import of Japanese cars and doubled taxes on the US and European cars, to protect the domestic industry, which was in its cradle, against any foreign competition. It also provided many facilities over three decades to the private sector. However, it is clear that government policies were temporary for each period of growth and changed periodically according to the degree of growth and the response of the local auto industry. We believe that this reduced the time between not owning the technology to imitating it and then innovating.

After the end of the Korean Civil War in 1953, the US bases and recruits increased on Korean soil, which increased the import of small transport vehicles for the army personnel of more than 300,000 soldiers. These transport vehicles needed local spare parts for cost reduction and quick supply, hence establishing a local industry sector for spare parts for transport vehicles for the US Army. In 1955, the Seong brothers set up a vehicle assembly plant (Jeep) for the US Willys Jeep Company. The assembled car was known at that time as Sibal (*lit. inception*). Its chassis or body shell was made from leftover US Army oil drums manually reshaped. Sibal became a relatively popular choice for Korean taxi firms. A replica of Sibal is on display in the National Museum of Korea. It is noteworthy that the main parts of the car were shipped from the US to be used as inputs in the Korean local production line (31-34).

Then the US and Japanese auto manufacturers established many of their production lines inside Korea, and these factories were wholly owned by foreign companies. In 1962, the Korean government issued the Automobile Industry Promotion Policy under which foreign automakers were barred from operating in Korea, except in joint ventures with local business entities holding a share in the local factory. Accordingly, foreign auto manufacturers searched for local partners. Three local car manufacturers (assembly lines for foreign products) appeared in addition to Seong, the agent of the US Jeep (10, 35).

The Saenera Company was established in 1962 as an assembly factory for the Japanese Nissan Company and Kia, which transformed its activity from the manufacture of bicycles since 1944 [exporting them to Hong Kong] to a factory for the assembly of the Mazda cars. Kia produced its Brisa S-1000 car, based on the Mazda 1000 model in 1974, the Brisa K-303, based on the Mazda 323 in 1981, and Pride, based on the Mazda-121 in 1987, the Concordia, based on the Mazda 626 in 1987, and the Potentia, based on the Mazda 929 in 1992. In 1965, the Shinjin Factory turned from spare parts for US army vehicles since 1954 to a car manufacturing factory and produced the Shin Sungho car in 1963, based on one of the well-known Nissan cars, the

Datsun Blue Bird 310. Asia Motors Company transformed its activity from assembling transport vehicles to assembling cars at the beginning of 1965. In 1970, it became an assembly line for the Italian Fiat 12M. As for the Hyundai Company, it was founded later in 1968 in cooperation with the US Ford Company and developed an assembly line for the Ford Cortina with a production capacity of 614 units per year (36) (37-39).

With the gradual change of government policies, the auto industry began to change from a foreign industry in Korea due to the abundance of low-waged labor and its proximity to points of sale inside Korea to foreign assembly factories with the participation of local ownership to keep pace with the regulations. Then another phase of government policies and plans aimed at raising content in the auto industry began. The plan to localize the auto industry was launched in 1965, aiming to reach 90% of local content within three years. Then again in 1973, the Korean government modified its policy regarding subsidizing production lines for car assembly from supporting locally developed Korean cars to cars manufactured in Korea and not assembly lines for foreign cars containing at least 95% local content to enhance independence from foreign companies. Accordingly, local auto companies began trying to develop their own products in the mid-1970s, but all attempts did not exceed copying, imitation, and simulation of foreign cars with a local tinge or a slight change to structures and designs in addition to the low quality of manufacturing. Attempts continued for more than a decade until innovation took its course (36, 40, 41).

With the export of Korean cars, Korea became known in the West as "Little Japan," as Japan was described before as "Britain of the East." Let's take a Korean car company that started in the 1960s as an example to get acquainted with the transformation in the Korean auto industry. Hyundai, for example, can be used as a model to illustrate the technological transformation in the auto industry from not having the technology to the ability to imitate and simulate and then to innovate in three to four decades.

As we pointed out earlier, Hyundai started in 1968 as a factory for an assembly line of the Ford Cortina, 618 of which were manufactured in the same year. Then, in response to government policies in 1973, it developed its first car, the Pony, which was a Korean-designed car. Still, most of its parts contained technologies licensed from foreign companies, for example, a license by Mitsubishi for the engine and transmission in exchange for a share in Hyundai. Although the local content in this car was 90%, the local content included textile, leather, iron, and parts that did not affect the main manufacture of the car and did not include the engine, transmission, and chassis (42, 43).

Hyundai used these technologies licensed by Mitsubishi to manufacture its Excel model launched in 1980. However, after the development of Hyundai, it exported its Pony and Excel cars and competed with the technology licensing company. Accordingly, Mitsubishi refused to grant a technological license to Hyundai's new engine with better specifications. Hyundai learned that there is no room for the newly developed generation of engines in this way and that its future growth may stop due to its inability to keep pace with progress in the auto industry. After other companies refused to license the technology, Hyundai knew that for sure (42, 44).

Hyundai was utterly dependent on imported technology licensed from abroad. From 1973 to 1985, the company signed 54 technological licenses with 26 companies in five countries (Britain, USA, Italy, Germany, and Japan). Even in the external and internal designs, the company resorted to foreign companies. It signed a training contract, transfer of expertise, and technological consultancy with the Italian company Italdesign, under which it sent five trainees to Italy upon signing the contract to stay for a year and a half to learn car designs under the supervision of the Italian company (45-47).

All Korean auto companies during the same period (1973-1985) resorted to the same method, but with a different foreign partner. Saugn signed nine technological licenses; Kia signed

14 technological licenses; and Daewoo signed 22 technological licenses. These companies opted for local R&D to develop the local auto industry and introduce developments and innovations that would make local companies not only get a share in the Korean domestic market, but also in the global market. Accordingly, the number of researchers in Hyundai Company jumped from 197 in 1975 to 2,247 in 1986, then to 3,890 in 1994, i.e., twenty times in less than two decades. As for R&D spending, the allocations increased from 1.1 billion Korean won in 1975 to 400 billion won in 1994, nearly 396 times in less than two decades (48, 49).

In the learning journey to move from imitation to innovation at Hyundai, the company required 14 months of testing (trial and error) the previous engine licensed from Japan's Mitsubishi to reach the next generation of engines (the Alpha engine). During these months, 11 engines exploded before engine number 12 operated successfully. Hyundai produced the first Korean engine in its Accent car in 1984, thanks to its developed R&D capabilities. With the Hyundai Accent, South Korea was the second country in Asia (after Japan) to have a local auto industry with national designs and capabilities. The company's share in the local market in 1970 was 19.2%, then 59% in 1976, after the production of the Pony model, which was also the first Korean car to be exported. Then, in 1979, the company's share in the local market became 74%, i.e., Hyundai acquired nearly three-quarters of the local market (48-50).

After introducing the Excel model, it became the best-selling foreign car in the USA in 1987, with more than a quarter of a million cars (263,610 cars per year) sold. In 1993, the Elantra became the best-selling car in Australia. In less than three decades, Hyundai's production rose from 614 units in 1968 to over one million in 1986. In 2021, the company became one of the five largest auto manufacturers worldwide. After the merger with Kia, the two companies' sales in the USA are now higher than those of all European companies combined. In 2010, the Sonata and Elantra models entered the list of the top ten best-selling cars in the world (48, 51).

In 2013, Hyundai's sales of US$ 117 billion exceeded those of Germany's BMW (US$ 98 billion), Japan's Honda (US$ 96 billion), and the French Peugeot (US$ 73 billion) (49).

~✲~

The above is not specific to Hyundai or the Korean auto industry alone. Still, this represents a transparent model for a specific company in a specific industry in Korea that shows how Korea moved in three to four decades from not having the technology and unable to copy and imitate to innovate, compete in global markets, and acquire larger market shares. The selection of Hyundai was due to its possession of nearly three-quarters of the domestic Korean market, as we mentioned earlier.

However, the reader would note a difference in the composition of companies, as the major industrial conglomerates (*Chaebols, lit. wealthy families*) control most of the Korean economy with the support of the Korean government. We will address this in future works. The critical issue here is that this model in corporate structures was initially adapted from the Japanese "Zibatsu" model (the major Japanese business conglomerates) developed by the Japanese in 1860 and planted in Korea during the thirty-five years of Japanese colonization of Korea after 1910 (38).

FROM IMITATION TO INNOVATION IN KOREAN SHIPBUILDING

Regardless of some historical successes in shipbuilding, which are difficult to link to the current situation, as there is no gradual, cumulative correlation over time, such as the case of the turtle ship (*Geobukseon*) built by the Koreans in 1592, equipped with 26 cannons and covered with an iron plate to repel the gunshots of the Japanese army during the Battle of Sacheon (52, 53), we found that the gradual beginning can be traced back to the 1960s to know how Korea has reached its position now in shipbuilding. In the 1970s, Korea produced low-quality replicas of Japanese ships. Then it moved to the ranks of top ship-manufacturing countries globally at the beginning of the third millennium.

In the 1960s, there were nine small shipyards for the manufacture and maintenance of steel ships from the remnants of the needs of the US Navy fleet, one-third of which left Korea in 1971. In 1967, the Korean government issued the Shipbuilding Promotion Law and established the first large shipbuilding port in Korea in 1970. It was owned by the Korean government and was called the Korean Shipbuilding and Engineering Corporation (KSEC) (54, 55).

This step was followed by a series of government measures to support this industry, including the establishment of large warehouses for the private sector at ports, the provision of financial credit and tax exemptions to the private sector, the most important of which was the large government ship purchase orders from the private sector. Then, the development plan for the shipbuilding industry was issued in 1973 by the Korean Ministry of Trade. In 1973-1974, Korea's production paralleled the production of Brazil and Taiwan, which was the beginning of entering the shipbuilding field. However, these ships were known as Japanese counterfeit ships of low quality in the West. For Korea to become the largest producer of ships globally despite its humble beginnings and lack of technology, Korean companies signed 159 technological licenses for shipbuilding from 1962 to 1987 (25 years) with a total value of US$ 117 million with foreign companies, including Hyundai's license from British companies Appledore and Scott Lightglow of US$ 1.7 million in 1972 in addition to a share of the company's future sales (0.5%).

The technological licensing contracts signed by Korean companies included a contract of US$ 281,000 with the German Naiereorm Company in 1979 for designs to build a ship with a capacity of 80 thousand tons, a contract worth about US$ 11 million with the Danish B&W Company in 1982 for designs of a ship with a capacity of 130,000 tons, and a contract worth US$ 100,000 with the Danish BWS Company in 1985 for designs for a ship with a capacity of 170,000 tons (54).

There was a turning point credited to the Korean government and entrepreneurs. During the global oil crisis in 1973 and the rise in oil prices, Japan and European countries began to reduce their production and stop the expansion of shipbuilding capacity. Korea exploited the crisis to raise production, purchase technologies at lower prices, and attract experts from companies with long experience in this field. The Hyundai Company established the Ulsan Shipyard in 1977 amid the economic oil crisis and the decline in ship consumption (56).

In figures, the production of European countries decreased from 12 million tons of ships in 1975 to 3 million tons in 1990, a quarter of the original production. The workforce in the shipbuilding industry in Europe decreased from 325,000 in 1975 to 155,000 in 1983. In Japan, it decreased from 150,000 in 1975 to 85,000 in 1983. In Korea, production increased from 410 thousand tons in 1975 to 3.44 million tons in 1990. Thus, the production of Korea (according to the production capacity) became more than the production of all of Europe. The top three Korean companies in production in 1990 were Hyundai with a capacity of 3 million tons, Daewoo with 2.754 million tons, and Samsung with 0.802 million tons (54).

In 2004, five of the Korean companies became among the top ten companies globally in the production of ships (according to construction capacity in tons). Hyundai Heavy Industries (HHI) production reached 144 million tons, Samsung Heavy Industries 91 million tons, and Daewoo Marine Engineering and Shipbuilding 88 million tons (57).

Back to the review of a Korean company in building a shipbuilding yard, the largest in Korea at the time, which exceeded the government shipbuilding dock established in 1970, KSEC, mentioned above. In June 1970, Hyundai obtained a license to build a shipyard in Ulsan, and construction began after obtaining financing of approximately US$ 60 million between 1972 and June 1974. Only US$ 10 million of this amount came from the company's resources and government support sources, as it obtained US$ 50 million in funding from a foreign bank (in Britain). The company relied on a group of British and Japanese companies to build the dock, such as the British Appledore Company, in the design of the dockyard, the Scott Lightglow Company for ship designs and operations, and the Japanese Kawasaki Company in shipbuilding. It also hired a group of foreign experts during the first three years, including the head of the port/dock, Dane Kart Schan. It produced the first two ships at the dock in Yilsan in 1974: the Atlantic Baron and the Atlantic Baroness. To raise the technological absorption capacity, the company sent 60 Korean trainees to Japanese and British companies (some of the companies referred to above), and then opened a local training center, where the trainees reached 3,636 trainees in 1975, and then more than 35,000 in 1990. During the first five years, it developed the first ship design with local capabilities with a capacity of 25 thousand tons in 1978, although the manufactured model was just a copy of the European and Japanese models, but it was the beginning before it bought technological licenses from many foreign companies.

Therefore, this chapter summarized Korea's efforts to transform the shipbuilding industry to its peak (54, 56, 58).

❦

FROM IMITATION TO INNOVATION IN KOREAN ELECTRONICS

Although Korean companies began dealing with electronics and semiconductors at the end of the 1950s and early 1960s to assemble radio and simple electric fans with techniques imported from Japan and the USA. During nearly fifty years, in 2019, Korea's market share of the world's semiconductor production reached 19% (about one-fifth of the global market share), outperforming all of Europe combined (10% market share) and Japan (10% market share) (59).

Analyzing Korea's superiority in electronics requires us to go back to the beginnings in imitation and copying that enabled Korea to innovate later. In 1958, the Korean LG Company began copying radio technologies and electric fans by establishing licensed assembly lines from the US and Japanese companies. In 1962, it began exporting its products to the US market at attractive prices due to the low labor cost. In 1991, LG sales amounted to US$ 11.7 billion and continued to rise in 1997 to

US$ 20.2 billion until the company's R&D spending reached US$ 1.4 billion in 1997 (60).

In 1975, the Korean Anam Company manufactured color television. It took electronics-related technologies from the Japanese Matsushita Company in return for 2% of sales revenue for color TV, 5% for CD players, and 7% for video cassette recorders. During the 1970s, the Dutch Philips Company gave a technological license to manufacture the CD player to ten Korean companies. In 1986, the Japanese Hitachi Company licensed the 1MB memory technology to Korean companies when it started manufacturing 4MB memory technologies, believing there would be no future competition from the Korean startups (60-63).

The market share of the Korean Samsung Company in 1984 was zero in the field of dynamic random-access memory (dynamic RAM or DRAM). In 1994, Samsung was one of the first companies to develop 256MB DRAM. In 1993, the global memory market represented 30% of the semiconductor market. Korea's share of the memory market was 29% of the global share, although Japan's share was still the highest, with 48%. However, in 1998, Korea managed to acquire a higher market share than Japan. An overview of the technological gap and the time lag of technology acquisition in the field of memory would show that Korea began producing 1MB memory in 1986, a year

and a half after it was invented by other countries. The technological gap took about a year to produce the 4MB memory in
1988 (invented in 1987), then the Korean performance began
to improve gradually, as it produced each of the 16MB memory
in 1990 (the same year it was invented) and the 64MB memory
in 1992. It even managed to be one of the first to innovate with
256MB memory (63, 64).

Samsung, which had the highest sales in Korea for memory
technology in the 1990s, had humble beginnings in entirely different fields. Lee Byung-Chul established it in 1938 as a store
selling noodles and dried fish, then moved to the wool industry in 1954, then to the insurance field in 1958, and started in
electronics in 1968, and black and white television in 1970. In
1978, it entered the field of manufacturing electric microwave
ovens, then air conditioners in 1980, and began manufacturing
liquid crystal display screens (LCDs) in 1995, and smartphones
in 1996 (65-67).

The goal of tracing the sequence of Samsung's transformation into various technological fields in electronics is to show that
the company began copying technologies imported from Japan,
the USA, and Europe in the late beginnings, and then managed
in 2010 to become the largest producer of LCDs worldwide
with a market share of 17.6%. The Korean LG Company won
the third-highest share in the LCD sales market, amounting to

11.8% in 2010. As for the smartphone market, Samsung became the largest producer of smartphones (according to production volume) in the fourth quarter of 2011. Samsung surpassed the two US companies Apple and Microsoft combined in the number of patents granted by the USPTO in 2014, with 4,936 patents compared to 4,832 for both US companies. The Korean company LG also obtained more than 2,000 patents from the USPTO, which exceeded those obtained by Apple (67-70).

Korean companies spent part of their profits on R&D, which contributed to multiplying the innovative content of Korean companies. LG spent approximately 6.9% of its sales on R&D in 1997. Despite the shift in Korea's sales in the electronics market in general from US$ 55 million in 1970 to US$ 22.2 billion in 1993, more than 400 times in less than two and a half decades (71), the content of copying and imitation is still the basis on which some Korean companies have relied. For example, the Korean company Samsung was fined US$ 539 million in 2018 for infringing the intellectual property of the US company Apple and copying its smartphone designs (68).

This chapter shows that Korea did not excel in manufacturing televisions, smartphones, air conditions, and other electronic technologies by working alone and reinventing what was invented by others, especially Japan and the USA. These technologies were imported from foreign countries in various ways (we explained part of them in this chapter), but it was not satisfied with the process of copying and imitation. Korea spent large sums on R&D until it reached innovation. In 2014, it reached the highest R&D spending worldwide as a GDP percentage (72).

References

1. L. Keun. (UNU-WIDER, Helsinki, Finland, 2009), vol. 2009.

2. The World Bank. (World Bank national accounts data, and OECD National Accounts data files, 2022).

3. M. Perry, Dynamic chart: World's ten largest economies, 1961 to 2017. *American Enterprise Institute (AEI),* (2018).

4. D. S. Yim, Korea's National Innovation System and the Science and Technology Policy. *STEPI,* (2014).

5. R. J. Myers. (Palgrave Macmillan, New York. , 2001).

6. C. V. Gilliland, JAPAN AND KOREA SINCE 1910. *Annual Publication of the Historical Society of Southern California* **11**, 47-57 (1920).

7. H.-K. Park, American Involvement in the Korean War. *The History Teacher* **16**, 249-263 (1983).

8. G. Warner, The Korean War. *International Affairs (Royal Institute of International Affairs 1944-)* **56**, 98-107 (1980).

9. J. D. Park. (Palgrave Macmillan, Cham, 2019).

10. K. S. Kim, THE KOREAN MIRACLE (1962-1980) REVISITED: MYTHS AND REALITIES IN STRATEGY AND DEVELOPMENT. *The Helen Kellogg Institue for International Studies,* (1991).

11. M. J. Seth, South Korea's Economic Development, 1948–1996. *Asian History,* (2017).

12. S.-H. Kim, Finance and Growth of the Korean Economy from 1960 to 2004. *Seoul Journal of Economics* **20**, (2007).

13. A. E. Fakir, in *Management of Technology Innovation and Value Creation.* pp. 275-292.

14. Y. Lee, *New Dawn: Republic of Korea and Syngman Rhee.* (CreateSpace Independent Publishing Platform, 2014).

15. B.-K. V. Kim, Ezra F., , *The Park Chung Hee Era: The Transformation of South Korea.* (Harvard University Press, 2013).

16. A. Roy, The Beginnings of Japan's Economic Hold over Colonial Korea, 1900-1919. *French Journal of Japanese Studies*, (2015).

17. M. Kimura, The Economics of Japanese Imperialism in Korea, 1910-1939. *The Economic History Review* **48**, 555-574 (1995).

18. Y. Chang, Colonization as Planned Changed: The Korean Case. *Modern Asian Studies* **5**, 161-186 (1971).

19. J. C. P. Mahlich, Werner *Innovation and Technology in Korea: Challenges of a Newly Advanced Economy.* (Physica-Verlag HD, 2007).

20. S. Haggard, B.-k. Kim, C.-i. Moon, The Transition to Export-led Growth in South Korea: 1954-1966. *The Journal of Asian Studies* **50**, 850-873 (1991).

21. R. Y. W. Westphal L.E., Pursell G. , *Sources of Technological Capability in South Korea.* (Palgrave Macmillan, London, 1984).

22. L. Kim, H. Lee, Patterns of technological change in a rapidly developing country: A synthesis. *Technovation* **6**, 261-276 (1987).

23. Y.-S. Kim. (1997).

24. S. Chung, paper presented at the Lessons From East Asia and the Global Financial Crisis, 2010.

25. D. Wang, X. Zhao, Z. Zhang, The Time Lags Effects of Innovation Input on Output in National Innovation Systems: The Case of China. *Discrete Dynamics in Nature and Society* **2016**, 1-12 (2016).

26. Y. Jaekyung, J. Byung Ho, C. Kangmin, in *2011 IEEE International Summer Conference of Asia Pacific Business Innovation and Technology Management.* (2011), pp. 221-225.

27. J. Jung, J. S. Mah, R&D Policies of Korea and Their Implications for Developing Countries. *Science, Technology and Society* **18**, 165-188 (2013).

28. A. O'Neill. (Statista.com, 2021).

29. J.-W. Lee, The Republic of Korea's Economic Growth and Catch-Up: Implications for the People's Republic of China. *ADBI Working Paper Series*, (2016).

30. United Nations University, "Monitoring and analysis of policies and public financing instruments conducive to higher levels of R&D investments The "POLICY MIX" Project," (United Nations University, , 2005).

31. L. Ho-jeong, in *Korea JoongAng Daily*. (2009).

32. H. S. Kim, My Car Modernity: What the U.S. Army Brought to South Korean Cinematic Imagination about Modern Mobility. *The Journal of Asian Studies* **75**, 63-85 (2016).

33. J.-M. Yang, T.-W. Kim, H.-O. Han, Understanding the economic development of Korea from a co-evolutionary perspective. *Journal of Asian Economics* **17**, 601-621 (2006).

34. S. Koo, A Study on Propriety of the Domestic Auto-makers Sports Car Developing: mainly on the Japanese auto-maker cases. 한국과학예술융합학회 **15**, 17-17 (2014).

35. C. K. Kim, C. H. Lee, Korean Automobile Industry. *The Motor Vehicle Industry in Asia: A Study of Ancillary Firm Development*, 286 (1983).

36. A. E. Green, South Korea's automobile industry: development and prospects. *Asian Survey* **32**, 411-428 (1992).

37. A. Jacobs, in *The Korean Automotive Industry, Volume 1*. (Springer, 2022), pp. 147-200.

38. A. H. Amsden, *Asia's Next Giant: South Korea and Late Industrialization*. (Oxford University Press, 1992).

39. V. Pattni, in *Top Gear*. (Top Gear, 2013).

40. J. I. Lee, J. S. Mah, The role of the government in the development of the automobile industry in Korea. *Progress in Development Studies* **17**, 229-244 (2017).

41. M. A. Yülek, K. H. Lee, J. Kim, D. Park, State capacity and the role of industrial policy in automobile industry: A comparative analysis of Turkey and South Korea. *Journal of Industry, Competition and Trade* **20**, 307-331 (2020).

42. A. Jacobs, in *The Korean Automotive Industry, Volume 1*. (Springer, 2022), pp. 239-269.

43. D. G. Southerton, *Hyundai and Kia motors: The early years and product development*. (Don Southerton, 2012).

44. Y.-s. Hyun, J. Lee, Can Hyundai go it alone? *Long Range Planning* **22**, 63-69 (1989).

45. Y.-s. Hyun, The road to the self-reliance new product development of Hyundai Motor Company. (2002).

46. S. Bencuya, The Miracle of Han River: Korean Government Policy and Design Management in the Motor Industry By Kyung-Won Chung. *Design Management Review* **25**, 63-73 (2014).

47. K. W. Chung, The miracle of Han River: Korean government policy and design management in the motor industry. *Design Management Journal (Former Series)* **4**, 41-47 (1993).

48. L. Kim, Crisis construction and organizational learning: Capability building in catching-up at Hyundai Motor. *Organization science* **9**, 506-521 (1998).

49. H. J. Jo, J. H. Jeong, C. Kim, Unpacking the 'black box'of a Korean big fast follower: Hyundai Motor Company's engineer-led production system. *Asian Journal of Technology Innovation* **24**, 53-77 (2016).

50. R. D. Lansbury, C.-S. Suh, S.-H. Kwon, *The global Korean motor industry: the Hyundai Motor Company's global strategy*. (Routledge, 2007).

51. L. Kraar, in *Fortune Magazine*. (Fortune, 1995).

52. W. G. Choi, *The Traditional Ships of Korea*. (Ewha Womans University Press, 2006), vol. 15.

53. J.-O. Park, J.-Y. Huh, J.-H. Huh, The Pedagogical Significance of Development of Turtle Ships: Focusing on the Scientific Perspectives involving Premier Ryu Seong-Ryong and Admiral Na Dae-Yong's Roles. *International Information Institute (Tokyo). Information* **20**, 5381-5389 (2017).

54. L. Bruno, S. Tenold, The basis for South Korea's ascent in the shipbuilding industry, 1970–1990. *The Mariner's Mirror* **97**, 201-217 (2011).

55. W. Shin, in *Shipbuilding and ship repair workers around the world.* (Amsterdam University Press, 2017), pp. 615-636.

56. J. Kang, S. Kim, H. Murphy, S. Tenold, Old Methods versus New: A Comparison of Very Large Crude Carrier Construction at Scott Lithgow and Hyundai Heavy Industries, 1970–1977. *The Mariner's Mirror* **101**, 426-457 (2015).

57. J. Park, T. Roh, Dynamics of Global Catch-up Process in Shipbuilding Industry: A Case of Hyundai Heavy Industries. 국제경영리뷰 **18**, 25-49 (2014).

58. J. Craggs, H. Murphy, R. Vaughan, A shipbuilding consultancy is born: The birth, growth and subsequent takeovers of A&P Appledore (International) Limited, and the A&P Group, 1971–2017. *International Journal of Maritime History* **30**, 106-130 (2018).

59. M. D. Platzer, J. F. Sargent, K. M. Sutter, in *R46581. Congressional Research Service. https://crsreports. congress. gov/product/details.* (2020).

60. J. W. Cyhn, *Technology transfer and international production: The development of the electronics industry in Korea.* (Edward Elgar Publishing, 2002).

61. M. Hobday, East Asian latecomer firms: learning the technology of electronics. *World development* **23**, 1171-1193 (1995).

62. M. Hobday, Export-led technology development in the four dragons: the case of electronics. *Development and Change* **25**, 333-361 (1994).

63. K. Kim, in *Proceedings of the Thirty-First Hawaii International Conference on System Sciences*. (IEEE, 1998), vol. 6, pp. 233-241.

64. M. Iansiti, J. West, David, i. Horii, *Technology integration: Turning great research into great products*. (Harvard Business School, 1997).

65. A. Jacobs, in *The Korean Automotive Industry, Volume 1*. (Springer, 2022), pp. 343-365.

66. Y. Zhou. (Nature Publishing Group, 2020).

67. R.-R. Park-Barjot, Samsung: An original case of knowledge transfer in economic organizations. *Entreprises et histoire*, 91-101 (2014).

68. G. Cain, *Samsung Rising: Inside the secretive company conquering Tech*. (Random House, 2020).

69. B. Crothers, in *Cnet*. (Cnet, 2012).

70. Statista Research Department. (Statista, 2011).

71. S. R. Kim, The Korean system of innovation and the semiconductor industry: A governance perspective1. *Industrial and Corporate Change* **7**, 275-309 (1998).

72. K. OECD., *OECD Science, Technology and Innovation Outlook 2018*. (OECD Publishing Paris, 2018).

WORD INDEX

www.ingramcontent.com/pod-product-compliance
Lightning Source LLC
Chambersburg PA
CBHW031930190326
41519CB00007B/477

* 9 7 8 1 7 3 4 6 2 8 7 8 4 *